T0275830

SpringerBriefs in Environmental Science

SpringerBriefs in Environmental Science present concise summaries of cutting-edge research and practical applications across a wide spectrum of environmental fields, with fast turnaround time to publication. Featuring compact volumes of 50 to 125 pages, the series covers a range of content from professional to academic. Monographs of new material are considered for the SpringerBriefs in Environmental Science series.

Typical topics might include: a timely report of state-of-the-art analytical techniques, a bridge between new research results, as published in journal articles and a contextual literature review, a snapshot of a hot or emerging topic, an in-depth case study or technical example, a presentation of core concepts that students must understand in order to make independent contributions, best practices or protocols to be followed, a series of short case studies/debates highlighting a specific angle.

SpringerBriefs in Environmental Science allow authors to present their ideas and readers to absorb them with minimal time investment. Both solicited and unsolicited manuscripts are considered for publication.

More information about this series at http://www.springer.com/series/8868

Mohamed Samer

Abatement Techniques for Reducing Emissions from Livestock Buildings

Mohamed Samer
Department of Agricultural Engineering,
Faculty of Agriculture
Cairo University
Giza
Egypt

ISSN 2191-5547 ISSN 2191-5555 (electronic)
SpringerBriefs in Environmental Science
ISBN 978-3-319-28837-6 ISBN 978-3-319-28838-3 (eBook)
DOI 10.1007/978-3-319-28838-3

Library of Congress Control Number: 2016931307

Printed on acid-free paper

This Springer imprint is published by SpringerNature
The registered company is Springer International Publishing AG Switzerland

Preface

This book provides useful information about emissions form livestock buildings and manure management. The theoretical considerations are described in this book as follows: Nitrogen cycle, nitrogen oxides, ammonia, hydrogen sulfide, methane, carbon dioxide, carbon monoxide, odors, dust, and aerosols. Furthermore, this book provides solutions on how to abate the emissions of gaseous pollutants from livestock buildings and manure management using effective emissions abatement techniques as follows: additives, covering manure storages, aerobic and anaerobic treatment, and dietary manipulation. On the other hand, dust emissions abatement techniques are discussed as follows: Spraying oil and water, oxidizing agents, ionization systems, aerodynamic dedusters, bioscrubbers, windbreak trees, and walls. Additionally, biofiltration for odor control is elucidated as follows: biofilter design and media of biofilter. The recent advancements in this field as well as the perspectives are discussed in this book.

Each chapter of the book provides precious and up-to-date knowledge from basics to apex, allowing readers to understand more deeply. This book will be very helpful for academics, scientists, scholars, researchers, undergraduate and graduate students worldwide who are specialized in engineering, environmental engineering, civil engineering, biosystems engineering, agricultural and biological engineering. Additionally, it will be very helpful for nongovernmental organizations (NGOs), universities, and research institutes and centers.

Mohamed Samer

Contents

About the Author

Dr. Mohamed Samer is an Associate Professor at the Department of Agricultural Engineering, Faculty of Agriculture, Cairo University, Egypt. Previously he was a Research Scientist at the Leibniz Institute for Agricultural Engineering Potsdam-Bornim (ATB) in Germany. He has held several positions at Cairo University: Teaching Assistant, Senior Teaching Assistant, and Assistant Professor. He got a long-term governmental scholarship to carry out his doctoral research abroad, from the Cultural Affairs and Scientific Missions Sector of the Ministry of Higher Education and the State for Scientific Research, Government of Egypt. He has been conferred the degree "Doctor scientiarum agrariarum" (Dr. sc. agr.) by the University of Hohenheim, Institute of Agricultural Engineering, Stuttgart, Germany. He speaks four languages fluently: English (TOEFL iBT 86 points, IELTS overall band score 6.5, and TOEFL CBT 233 points), German (two International Certificates: ZD and ZMP from Goethe-Institute, 15 levels), French (student at French schools), and Arabic (his native language). He has more than 70 publications, 33 thereof are peer-reviewed articles published in international journals with impact factor (e.g., Building and Environment, Biosystems Engineering, Renewable Energy, Energy and Buildings). He has attended more than 25 international conferences and workshops, where he participated in organizing eight international conferences and exhibitions. Additionally, he has taught 15 undergraduate and postgraduate courses in his career. He has previously led three research projects and currently leading three research projects as a principal investigator (PI). Besides, he has participated in several other projects as a Co-PI and as a research team member. He has attended more than 20 professional training courses. He has visited several countries: USA, Canada, UK, Germany, France, Austria, Switzerland, The Netherlands, Belgium, Poland and Greece. Furthermore, he is a member of 11 international scientific societies (e.g., ISIAQ, A&WMA, AEESP, ASABE, CSBE, CIGR Emission Network). He serves as a peer reviewer for more than 20 scientific journals, where he reviewed more than 80 research papers. He has supervised 3 Ph.D. students, 5 M.Sc. students, 37 B.Sc. students, 2 trainees, 15 technicians, and 10 staff members. His main research and teaching interests are as follows: Biosystems engineering, environmental engineering, renewable energy, bioenergy, biofuels (biodiesel, bioethanol, biogas), pyrolysis (biochar, bio-oil, syngas), air quality engineering, natural and mechanical ventilation, gaseous emissions (greenhouse gases, ammonia), odor emissions, dust emissions, emissions abatement techniques, emissions inventory, biofilters and biofiltration, agricultural waste management, slurry treatment, manure management, wastewater treatment, bioreactor design, biogas production, agricultural buildings, green buildings, heating, ventilation and air-conditioning (HVAC), microclimate in livestock housing, planning and designing livestock buildings, structures and construction materials, mathematical modeling, precision livestock farming, information technology, and expert systems.

Abstract

The air pollutant emissions from agriculture have negative environmental impact and pertinent political importance (Kyoto Protocol and Gothenburg Protocol). Animal production is a major source of atmospheric pollutants, such as: methane (CH_4), nitrogen oxides (NO_x), carbon dioxide (CO_2), and ammonia (NH_3). Methane, nitrogen oxides, and carbon dioxide are greenhouse gases (GHGs) that contribute to the global warming and, therefore, the climate change. Ammonia is responsible for eutrophication and soil acidification. This study elucidates and illustrates the theoretical background of the development, release, and spreading of NH_3, CH_4, NO_x, hydrogen sulfide (H_2S), dust, and odors in livestock buildings. Subsequently, the emissions abatement techniques for reducing air pollutants (e.g., GHGs, NH_3, H_2S, dust, odors) emissions from livestock buildings have been clarified and discussed. The emissions abatement techniques presented in this study focuses on the manure handling especially inside livestock buildings, dust mitigation, biofiltration for pollutants and odor control, biofilter design and operating parameters, and bioscrubbers. Furthermore, this study identifies future scientific research priorities for developing emissions inventories, emissions abatement techniques, and mitigation strategies in order to improve and sustain livestock production to be in line with the climate change adaptation.

Keywords Abatement techniques · Ammonia · Animal buildings · Biofilters · Dust · Emissions · Greenhouse gases · Livestock housing · Mitigation strategies · Odor

Chapter 1
Introduction

The United Nations Framework Convention on Climate Change (UNFCCC) adopted the Kyoto Protocol (Protocol that set binding obligations on the industrialized countries to reduce their emissions of greenhouse gases), where the Kyoto Protocol aimed at achieving the stabilization of greenhouse gas concentrations in the atmosphere at a level that would prevent dangerous anthropogenic interference with the climate system. Under this protocol, several countries committed themselves to a reduction of the greenhouse gases. The protocol was adopted on December 11, 1997, in Kyoto, Japan, and entered into force on February 16, 2005. As of September 2011, 191 states have signed and ratified the protocol; the members exert efforts to reduce the greenhouse gases (GHG) emissions.

Following the adoption of the United Nations Economic Commission for Europe (UNECE) Gothenburg Protocol (Protocol to the 1979 convention on long-range transboundary air pollution, United Nations Economic Commissions for Europe (UNECE), Geneva), the members struggle to achieve significant reduction in national ammonia (NH_3) emissions. The Gothenburg Protocol is a multi-pollutant protocol designed to reduce acidification, eutrophication, and ground-level ozone by setting emissions ceilings for several pollutants, where ammonia is one of them.

Agriculture, with its two main sectors plant and animal production, is one of the main sources of greenhouse gases' emissions and the main source of ammonia emissions. The reduction of emissions of air pollutants is subject of international conventions, which include reporting of emissions in accordance with guidelines. Thus, reducing greenhouse gases and ammonia emissions from the agricultural sector is essential. Consequently, it is crucial to develop mitigation strategies to reduce GHGs and ammonia, which should be preceded by inventorying these emissions. There are two emissions inventory guides: the air pollutant emission inventory guidebook of the European Environment Agency (EEA) and the Cooperative programme for monitoring and evaluation of the long-range transmission of air pollutants in Europe (EMEP), and the guidelines of the Intergovernmental Panel on Climate Change (IPCC).

M. Samer, *Abatement Techniques for Reducing Emissions from Livestock Buildings*, SpringerBriefs in Environmental Science,
DOI 10.1007/978-3-319-28838-3_1

Animal production contributes with 65 % of the global anthropogenic nitrous oxide (N_2O) emissions and account for 75 to 80 % of the emissions from agriculture. Enteric fermentation and manure management account for 35 to 40 % of the total anthropogenic methane (CH_4) emissions and 80 % of CH_4 release from agriculture (FAO 2006). CH_4 and N_2O are greenhouse gases (GHG) with global warming potentials (GWP) of 23 and 296 times that of carbon dioxide (CO_2), respectively (IPCC 2007a). About 94 % of global anthropogenic emissions of ammonia (NH_3) to the atmosphere originate from the agricultural sector of which close to 64 % is associated with livestock management (FAO 2006). Around 75 % of ammonia (NH_3) emissions come from livestock production (Reinhardt-Hanisch 2008). Excessive levels of NH_3 emissions contribute to eutrophication and acidification of water, soils, and ecosystems (Schuurkes and Mosello 1988). In addition to the global warming potential of the greenhouse gases, ammonia emissions contribute to global warming when the ammonia is converted into nitrous oxide (Berg 1999; Sommer et al. 2000). On the other hand, hydrogen sulfide (H_2S) is classified as poison and causes death at 1000 mg/L, where, in most cases, death occurs when opening the manure storages for manure removal and, therefore, all team members must leave the area for a while. Regarding odors and dust/aerosols, there are increasing difficulties which are expected in the near future where nuisance and health effects are in question.

The main objective of a number of current research projects is the evaluation of the consequences of predicted climate change on different aspects on the environment and human life. These studies base their estimations on the current predictions of GHG emissions and temperature rise reported in the literature that will determine the extent of the consequences (Kuczynski et al. 2011). The assessment of climate change requires a global perspective and a very long time horizon that covers periods of at least a century. As the exact knowledge of future anthropogenic GHG emissions is impossible, emissions scenarios become a major tool for the analysis of potential long-range developments. According to IPCC (2007b) scenarios are a plausible and often simplified description of how the future may develop, based on a coherent set of assumptions about driving forces and key relationships. Scenarios are images of the future, or alternative futures. They are neither predictions nor forecasts. Rather, each scenario is one alternative image of how the future might unfold. Emissions scenarios are a central component of any assessment of climate change. Scenarios facilitate the assessment of future developments in complex systems that are either inherently unpredictable or have high scientific uncertainties.

The large difference between predictions of the different scenarios indicates the complexity involved in making such predictions and the large amount of uncertainty inherent in climate change models (Kuczynski et al. 2011). Even though a general trend was presented in the IPCC report (2007c): (a) for the next two decades, a warming of about 0.2 °C per decade is projected for a range of emission scenarios; (b) even if activities having an impact on the balance between energy entering and exiting the planetary system were reduced and held constant at year 2000 levels, a further warming trend would occur over the next two decades at a rate of about 0.1 °C per decade, due mainly to the slow dynamic response of the oceans; (c) continued GHG emissions at or above current rates would cause further warming and induce many changes in the global climate system during the twenty first century that would very likely be larger than those observed during the twentieth century (IPCC 2007c). Regarding the geographical distribution of the climate change, projected warming in the twenty first century shows scenario-independent geographical patterns similar to those observed over the past several decades (Kuczynski et al. 2011). Furthermore, the Special Report on Emissions Scenarios (SRES 2000) projected global average surface warming and developed relevant emissions scenarios.

A detailed knowledge of the processes of GHGs and NH_3 mass transfer from the manure and transport to the free atmosphere will contribute to development of emissions abatement techniques and housing designs that will contribute to the reduction of gaseous emissions to the atmosphere (Sommer et al. 2006). Carew (2010) stated that further research is needed to understand the factors limiting livestock producers adopting emissions abatement techniques and mitigating strategies to reduce emissions since a whole-farm system approach can provide a modeling framework to evaluate the feasibility and cost-effectiveness of abatement measures. While Samer (2015) studied the mitigation strategies for reducing the emissions of GHGs from manure, the present study focuses on the abatement techniques for reducing the emissions of air pollutants from livestock buildings.

Chapter 2
Theoretical Considerations

Emission can be defined as the release of pollutants from the source to the environment. Transmission can be defined as the distribution and conversion of pollutants during the atmospheric transport. Immission can be defined as the concentration and deposition of pollutants with impact on the places and creatures exposed (Fig. 2.1).

Livestock housing is a major source of harmful gases, e.g., CH_4, NH_3, CO_2, and N_2O (Zhang et al. 2011). Gaseous emissions measurements in livestock buildings are important as these pollutants may affect the health of farmers and the surrounding environment. Emission monitoring enables judgments on the effectiveness of mitigation strategies and controls on emission targets (Ngwabie et al. 2009), as well as the health and well-being of the animals. Table 2.1 shows the characteristics, global warming potentials, maximum indoor gas concentrations, and physiological effects of CO_2, NH_3, CO, H_2S, CH_4, and N_2O.

Manure management, inside and outside of livestock buildings, is responsible of emitting several gases. Manure is a mixture of solid and liquid animal excreta (feces and urine) collected from animal buildings, whereas dung is solid animal excreta, i.e., feces. Slurry is a mixture of scraped manure and flushing water and is collected from animal buildings. Hence, slurry is a mixture of manure and water. On the other hand, litter is animal excreta and bedding material collected from animal buildings (Samer 2011a). Livestock excreta stored in manure stores, in housing, in beef feedlots, or cattle hardstandings are the most important sources of GHGs and NH_3 in the atmosphere. The storage of dry manure produces large emissions of N_2O, while storage of liquid manure produces large emissions of CH_4 (Janzen et al. 2008). Inventories have shown that stored animal manure, animal housing, and exercise areas account for about 69–80 % of the total emission of NH_3 in Europe (ECETOC 1994; Hutchings et al. 2001). Most of CO_2 is formed by the animals and exhaled by respiration. It can also be part of exhaust gases of heating systems being

M. Samer, *Abatement Techniques for Reducing Emissions from Livestock Buildings*, SpringerBriefs in Environmental Science,
DOI 10.1007/978-3-319-28838-3_2

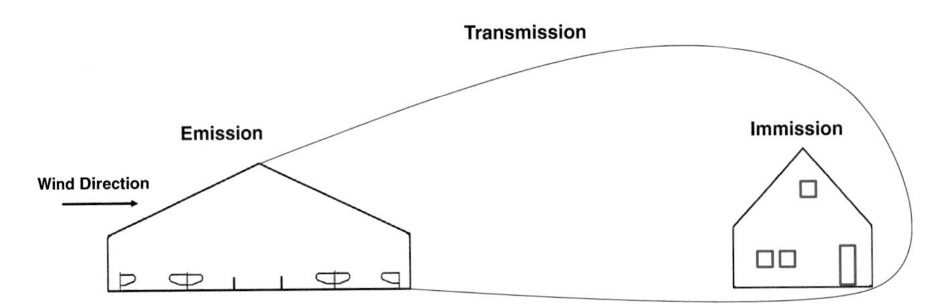

Fig. 2.1 Emission of pollutants from livestock buildings, transmission, and immission into neighborhood (*Amended, redrawn and adopted from* KTBL 2006)

released in the barns. Additionally, a certain portion of CO_2 is released by the manure. The released CO_2 from urine and dung in stored manure is less than 5 % of the amount produced by respiration (Schneider 1988; Aarnink et al. 1992). Low concentrations of N_2O can be measured in dairy barns with liquid manure systems, where daily or frequent manure removal into external storage tanks is applied and this does not constitute a major source of N_2O (Ngwabie et al. 2009).

Several algorithms for calculating methane and nitrous oxide emissions from manure management were developed. The biogenic emissions of CH_4 and N_2O from animal manure are stimulated by the degradation of volatile solids (VS) which serve as energy source and a sink for atmospheric oxygen. Algorithms which link carbon and nitrogen turnover in a dynamic prediction of CH_4 and N_2O emissions during handling and use of liquid manure were developed and include a sub-model for CH_4 emissions during storage relates CH_4 emissions to VS, temperature, and storage time, and estimates the reduction in VS; and a second sub-model estimates N_2O emissions from field-applied slurry as a function of VS, slurry N, and soil water potential, but emissions are estimated using default emission factors. Anaerobic digestion of slurry and organic waste produces CH_4 at the expense of VS. Accordingly, these models predicted a 90 % reduction of CH_4 emissions from outside stores with digested slurry, and a >50 % reduction of N_2O emissions after spring application of digested as opposed to untreated slurry. Additionally, simple algorithms to account for ambient climatic conditions may significantly improve the prediction of CH_4 and N_2O emissions from animal manure. Besides, several algorithms were developed for determining ammonia emission from buildings housing cattle and pigs and from manure stores (Sommer et al. 2004, 2006).

The factors-of-influence (FOI) that strongly influence the dispersion of NH_3 are NH_3 mass flow, internal and external temperatures, mean and turbulent wind components in horizontal and vertical directions, atmospheric stability, and exhaust air height where the continuous measurement of NH_3 remains a challenging and costly enterprise, in terms of capital investment, running costs or both (Von Bobrutzki et al. 2010; Von Bobrutzki et al. 2011). The determination of

Table 2.1 Properties, GWP, maximum gas concentrations, and physiological effects of some noxious gases (CIGR 1984, 1994, 1999; FAO 2006; IPCC 2007a; UNFCCC 2014)

Gas	Chemical formula	Lighter than air	Odor	GWP (100 years)	Class	Maximal indoor concentration	Comments
Methane	CH_4	Yes	Odorless	21	Asphyxiant flammable	–	Concentrations between 5000 and 15,000 ppm are explosive, several explosions have occurred due to ignition of methane-rich air in poorly ventilated livestock buildings
Nitrous oxide	N_2O	No	Slightly sweet odor	310	Anesthetic	3 ppm	Colorless and nonflammable gas, with a slightly sweet odor. Known as "laughing gas" due to the euphoric effects of inhaling it
Ammonia	NH_3	Yes	Sharp and pungent	Contributes to global warming only when converted to N_2O	Irritant	20 ppm	Irritation of eyes and throat at low concentrations; asphyxiating, could be fatal at high concentrations with 30–40 min exposure
Hydrogen sulfide	H_2S	No	Rotten eggs	NA	Poison	0.5 ppm (shortly 5 ppm during manure removal)	Headaches, dizziness at 200 mg/L for 60 min; nausea, excitement, insomnia at 500 mg/L for 30 min; unconsciousness, death at 1000 mg/L
Hydrogen cyanide	HCN	Yes	Bitter almond	NA	Poison flammable	10 ppm	A very toxic and explosive gas; released together with H_2S during mixing of manure
Carbon dioxide	CO_2	No	Odorless	1	Asphyxiant	3000 ppm	$<$20,000 mg/L is in the safe level; increased breathing, drowsiness, and headaches as concentration increases; could be fatal at 300,000 mg/L for 30 min
Carbon monoxide	CO	Yes	Odorless	3	Poison	10 ppm	Colorless, odorless and tasteless gas

emission mass flow is necessary not only to compute dispersion but also to develop mitigation strategies. While husbandry, dunging, and feeding influence the ammonia emission, likewise for both forced ventilation and natural ventilation, the building envelope including ventilation openings (design and control) and the outside climatic conditions are the dominant influencing factors (Samer and Abuarab 2014; Samer 2012b; Samer et al. 2011c).

The highest average ammonia emission coincides with higher environmental temperature. The gaseous emissions from naturally ventilated cattle buildings significantly increase with air temperature (Morsing et al. 2008; Adviento-Borbe et al. 2010; Pereira et al. 2011). Low emission values can only be achieved by reducing the emission source surfaces, decreasing temperature and air velocity near the source, and minimizing volumetric airflow rates throughout the livestock buildings (Adviento-Borbe et al. 2010; Bjorneberg et al. 2009; Blanes-Vidal et al. 2007; Gay et al. 2003). The drawing-off emission flux of harmful gases from a naturally ventilated building is dependent on wind velocity (speed and direction) and turbulence fields inside and over the building envelope; therewith the emission mass flow is highly variable and difficult to estimate (Ngwabie et al. 2009; Van Buggenhout et al. 2009; Hellickson and Walker 1983). The effects on gas emissions are as a consequence of changing airflow patterns and different types of flow in the boundary layer between the slurry and ventilation air.

In order to quantify the gaseous emissions, the tracer gas technique was developed (Samer et al. 2014a). The tracer gas technique is one of the approaches used for quantifying gaseous emissions and estimating ventilation rates in naturally ventilated buildings which including the measurement of infiltration, air exchange, and the dispersion of pollutants (Samer et al. 2011d). This technique implements tracer gases such as CO_2, SF_6, and Krypton 85 for measuring the ventilation rates and to calculate the emission streams. The emission mass flow from the livestock building is then the product of both the concentration difference between emitted and fresh air and the ventilation rate. The gaseous concentrations varied in time and place inside the investigated barn (Samer et al. 2012d; Samer et al. 2011e).

The ventilation rate and the gaseous emissions from a naturally ventilated livestock building are dependent on wind velocity. In order to investigate the distribution of air temperature and gaseous concentrations throughout the different zones of the building and to achieve an efficient control of the bio-responses, continuous monitoring and controlling of the micro-environment to variations of air velocity (direction and speed) inside the building is required. Therefore, the air profiles should be investigated and airflows should be analyzed through the zones of the building (Samer et al. 2011a; Berckmans and Vranken 2006). Large fluctuations occur in ventilation rates estimated using the combined effects of wind pressure and temperature difference forces, owing to large fluctuations in the wind velocity. The fluctuations of wind velocity (direction and speed) negatively affect the estimation of ventilation rates and then the gaseous emissions (Samer et al. 2014a; Samer et al. 2011d, e). Therefore, the airflow profiles should be investigated and airflows should be analyzed in livestock barns.

Investigating the airflow profiles inside a livestock building is important to determine the air and pollutants' distribution (Samer 2012a). The air movement can be characterized by velocity measurements and observation by visualization of the air flow pattern by smoke, where these images can be recorded by video camera and analyzed by computer image analysis. Via laser-light-sheet technique the air flow is made visible by smoke particles. The snapshot is digitally recorded and average images are calculated afterward. Airflow patterns in animal buildings influence the distribution of air temperature, gas concentrations, and the release of gases from manure. Air velocity measurements have been used for airflow pattern measurements.

Odors and gases emitted from animal houses are strongly related to airflows (Morsing et al. 2008). Sun et al. (2002) developed computational fluid dynamics (CFD) models to simulate air velocity and ammonia distribution within a hog building. Snell et al. (2003) stated that ventilation rate could be explained by the climatic values (wind velocity, wind direction, temperature, and relative air humidity), where the wind velocity is of central importance for the ventilation. Bartzanas et al. (2007) stated that air velocity measurements incarnate the corner stone for airflow analysis in rural buildings. Bjerg and Sørensen (2008) carried out numerical analysis and mentioned that to fulfill modern demands of airflow in livestock buildings, several procedures—which requires air velocity measurements—should be implemented, and they are determining air velocity at animal level, limiting air velocity in the animal occupied zone, homogenizing air velocity distribution in the entire barn, determining whether air velocity distribution inside and close to the inlet is similar, investigating air velocity profiles and turbulences, homogenizing air velocity direction throughout the entire barn, and reducing air velocity at floor level at high ventilation rate without increasing the pressure drop over the inlet.

2.1 Nitrogen Cycle

A summary of the major remodeling processes in the terrestrial nitrogen cycle is shown in Fig. 2.2. The individual conversion processes are marked with numbers. The main processes (Fig. 2.2) in the nitrogen cycle are nitrogen assimilation (no. 1 and 2); synthesis of endogenous proteins (no. 3); ammonification (no. 4, 5, and 6); direct deposit into soil (no. 7); emission (no. 8 and 9); transmission, deposition, and immission (no. 10 and 24); immobilization (no. 11); ammonium fixation and release (no. 12 and 13); nitrification (no. 14 and 15); leaching and capillary rise (no. 16 and 17); assimilatory nitrate reduction (no. 18 and 19); denitrification (no. 18, 20, 21 and 22); photochemical oxidation and chemical fixation (no. 23); and biological nitrogen fixation (no. 25).

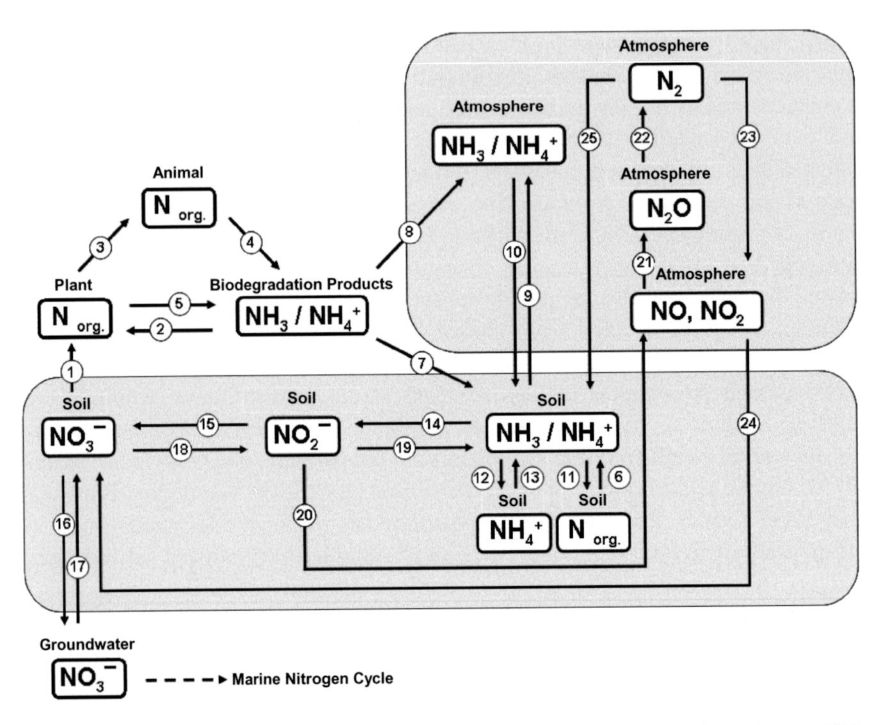

Fig. 2.2 Nitrogen cycle in the environment (*Amended, translated and adopted from* Jensen 1974; Lehninger 1977; Schilling et al. 1989; Reinhardt-Hanisch 2008)

2.2 Nitrogen Oxides

Nitrogen oxides (NO_x) consist of nitric oxide (NO), nitrogen dioxide (NO_2), and nitrous oxide (N_2O) and are formed when nitrogen (N_2) combines with oxygen (O_2). Nitrous oxide and nitric oxide can be released through nitrification and denitrification processes. Nitrification is the bacterial oxidation from nitrite to nitrate under aerobic conditions, as follows:

$$NH_4{}^+ \rightarrow NH_2OH \rightarrow N_2O(\uparrow) \rightarrow NO_2^- \rightarrow NO_3{}^-$$

Denitrification is the reduction of nitrite/nitrate to N_2 under anaerobic conditions, as follows:

$$NO_3^- \rightarrow NO_2^- \rightarrow NO \rightarrow N_2O\ (\uparrow) \rightarrow N_2$$

If the above-mentioned processes do not result in a fully conversion of the N-bonds because of suboptimal conditions, N_2O can be released.

Both nitrite and nitrate bacteria are carbon-autotrophic bacteria. Under strictly aerobic conditions, the bacteria use the released energy during nitrification for

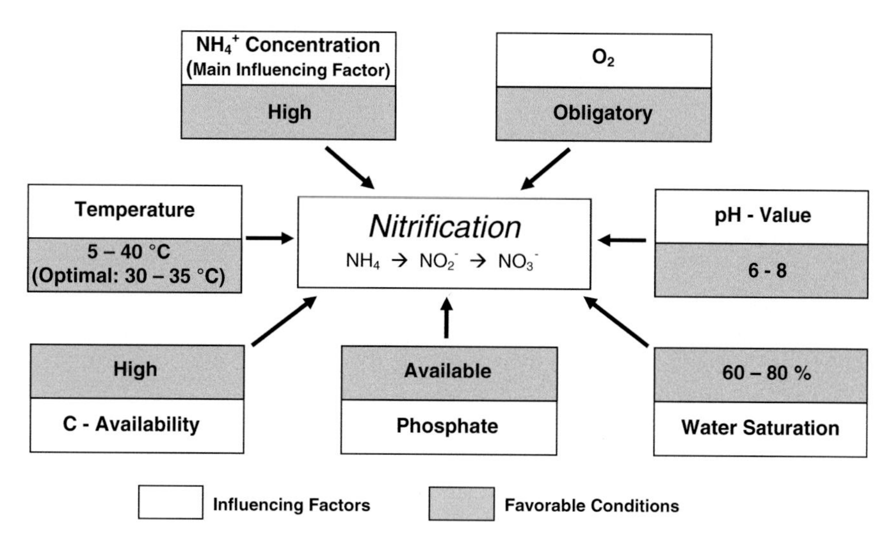

Fig. 2.3 Influencing factors of autotrophic nitrification (*Amended, translated and adopted from* Amon 1998; Reinhardt-Hanisch 2008)

assimilation of carbon dioxide (chemosynthesis) and therefore do not require organic carbon. The optimum pH is between pH 6 and 8. The influencing factors of nitrification and denitrification (Figs. 2.3 and 2.4) were illustrated by Amon (1998) and further amended by Reinhardt-Hanisch (2008). Under adverse conditions N_2O and NO can be released during nitrification (Fig. 2.5).

Under adverse conditions, such as increasing nitrate or nitrite concentrations (electron acceptors), increasing oxygen concentration, decreasing concentration of carbon, decreasing pH, decreasing temperature, and decreasing N_2O reductase activity, incomplete denitrification occurs and N_2O and NO are released (Fig. 2.5).

Nitrogen has enormous environmental effects, where humans have radically changed natural supplies of nitrates and nitrites. The main cause of the addition of nitrates and nitrites is the extensive use of fertilizers. Combustion processes can also increase the nitrate and nitrite supplies, due to the emission of nitrogen oxides that can be converted into nitrates and nitrites in the environment. Nitrates and nitrites also form during chemical production and they are used as food conservers. This causes groundwater and surface water nitrogen concentration, and nitrogen in food to increase greatly. The addition of nitrogen bonds in the environment has various effects. First, it can change the composition of species due to susceptibility of certain organisms to the consequences of nitrogen compounds. Second, mainly nitrite may cause various health effects in humans and animals. Food that is rich in nitrogen compounds can cause the oxygen transport of the blood to decrease, which can have serious consequences for cattle. High nitrogen uptake can cause problems in the thyroid gland and it can lead to vitamin A shortages. In the animal stomach and intestines, nitrates can form nitroamines, dangerously carcinogenic compounds.

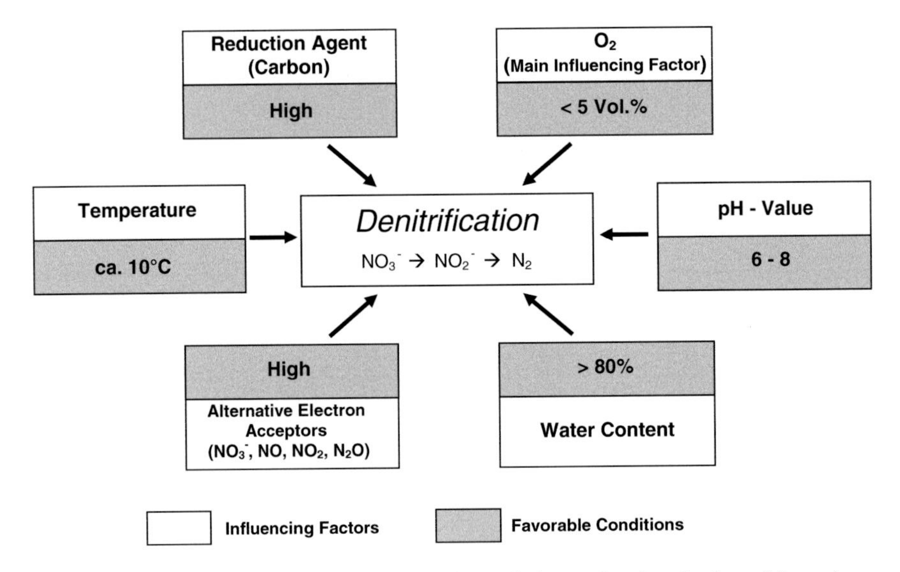

Fig. 2.4 Influencing factors of denitrification (*Amended, translated and adopted from* Amon 1998; Reinhardt-Hanisch 2008)

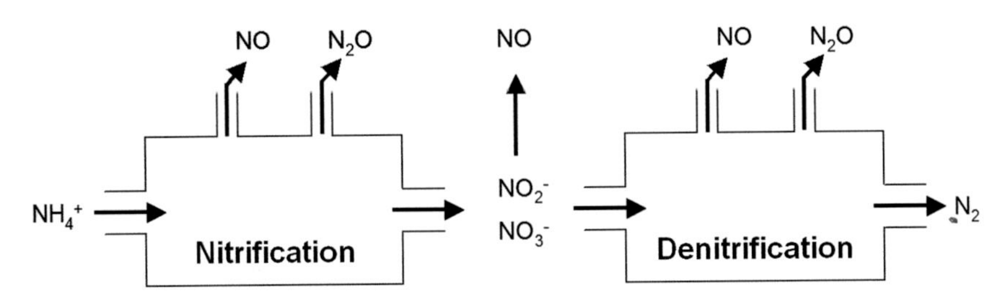

Fig. 2.5 Formation and release of N_2O and NO by nitrification and denitrification (*Amended by* Reinhardt-Hanisch 2008 *adopted from* Colbeck and Mackenzie 1994)

2.3 Ammonia

Ammonia, a colorless and highly water-soluble gas, is primarily an irritant and has been known to create health problems for animals in confinement building. Irritations of the eyes and respiratory tract are common problems from prolonged exposure to this gas. Ammonia can be detected by humans at levels as low as 5 mg/L and can reach levels of 200 mg/L in poorly ventilated buildings. Recently, the most common complaints against animal producers involve odor, and the primary component of odor is ammonia. Furthermore, very high levels of ammonia concentrations, such as 2500 ppm, may even be (rapidly) fatal. In several countries the labor inspectorate has established standards for ammonia concentrations, the so-called threshold values that should not be exceeded. In many countries, the threshold limit is 25 ppm (time weighted) for an 8 h working day for staff and for the living environment for livestock, while a higher limit is often applied for short-term exposures, e.g., 35 ppm over 15 min in England. However, sometimes the limit is stricter, e.g., 10 ppm for stockmen in Sweden. Shorter working days may allow higher threshold values, but little is known about the long-term effects of gaseous ammonia in the working environment. However, lower concentrations are always preferable to higher concentrations, both for workers and livestock. Ammonia emissions are expressed in mg NH_3 m^{-2} h^{-1}; however, ammonia emission factor is expressed in kg NH_3 per place and year where this value is 4.86 for tie-stalls and 14.57 for freestalls.

The effect of ammonia on the environment due to acidification and eutrophication can be severe. It is associated with soil acidification processes and eutrophication. Ammonia and its chemical combinations (NH_x) are important components of acidification in addition to sulfur compounds (SO_x), nitrogen oxides (NO_y), and volatile organic compounds (VOC). Ammonia is released from manure and urine, and is most noticeable during storage and decomposition. Formation of ammonia is induced by catalytic breakdown of urea as follows (Reinhardt-Hanisch 2008):

$$CO\,(NH_2)_2 + H_2O \overset{urease}{\longrightarrow} 2NH_3 + CO_2$$

Regarding the $NH_3 \leftrightarrow NH_4^+$ equilibrium in liquid, the higher the temperature and the higher the pH value of manure, the more the NH_3 production, i.e., the higher the emission potential. Furthermore, NH_3 release is based on mass transfer from NH_3 solved in the liquid to NH_3 in the air (Fig. 2.6). The main accelerating factors are high temperature, high air velocity, high turbulence of the air stream, and large size of the emitting surface. Seethapathy et al. (2008) stated that in winter lower atmospheric NH_3 concentrations occur due to the reduced volatility, lower temperatures, and the generally higher relative humidity.

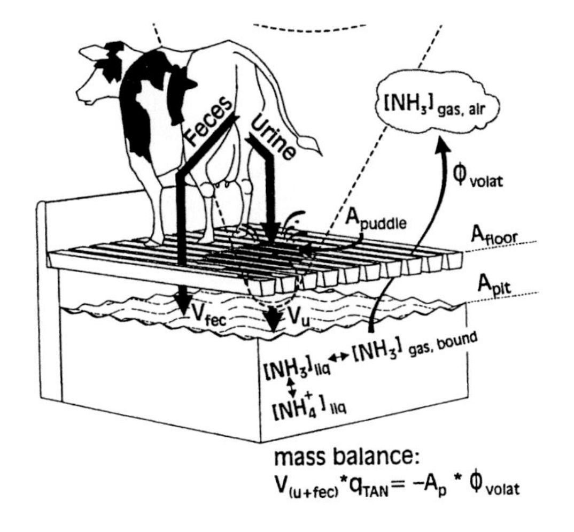

Fig. 2.6 Ammonia formation and release (Monteny 2000)

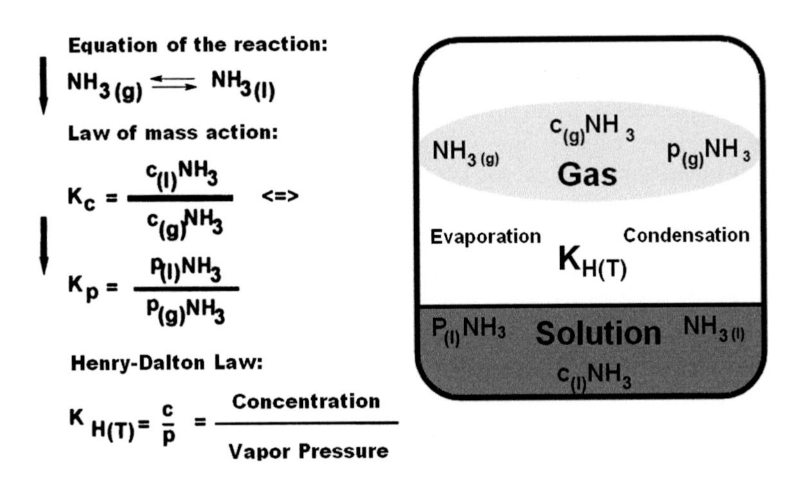

Fig. 2.7 Ammonia mass transfer from liquid to gas (*Amended, translated and adopted from* Hartung 1995)

The difference between the partial pressure of ammonia in the liquid and the partial pressure of the gaseous ammonia in the air of the boundary layer above the contact surface is the partial pressure difference. In a closed system (Fig. 2.7), a dynamic balance occurs between the amount of dissolved ammonia ($NH_{3(l)}$) and the amount of gaseous ammonia ($NH_{3(g)}$) is established. Figure 2.7 shows the ammonia mass transfer from liquid to gas, where in practice, the liquid is the manure and/or the contaminated surfaces inside the livestock building and the gas is the indoor air of the building.

Ammonia is released by excrements under special biochemical (pH value, temperature, and microorganisms) and physiological (species, age, feeding, and animal activity) conditions. NH_3 is released from different places, e.g., contaminated laying and walking areas, dirty animals, and manure stored inside the barn. NH_3 is released into indoor air in a certain concentration (g m^{-3}) which depends on the airflow, partial pressure, surface areas, manure handling system and housing system, and design (Fig. 2.8). Depending on the volumetric airflow rate, i.e., ventilation rate (m^3 h^{-1}), NH_3 is emitted in the exhaust air to outside of the barn in a certain mass flow emission rate (g h^{-1}).

Ammonia is only transported over short distances (transmission) in the atmosphere, and is deposited close to the emission source as dry deposition and is then entered into the soil. In the form of ammonium and various intermediates (Fig. 2.9), the nitrogen can be transported over long distances before it is usually deposited as wet deposition and entered into the soil (Dämmgen and Erisman 2006). An overview of the emission, dispersion, vertical and horizontal transport, and chemical reactions and deposition of ammonia and ammonium is shown in Fig. 2.9.

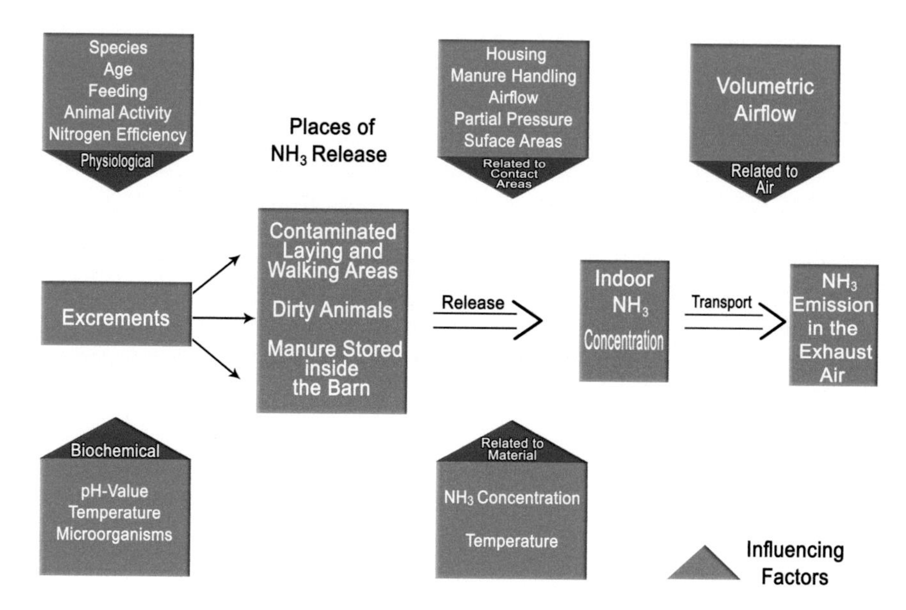

Fig. 2.8 Development, release, and spreading of ammonia inside the barn (*Amended, translated, redrawn and adopted from* Keck 1997)

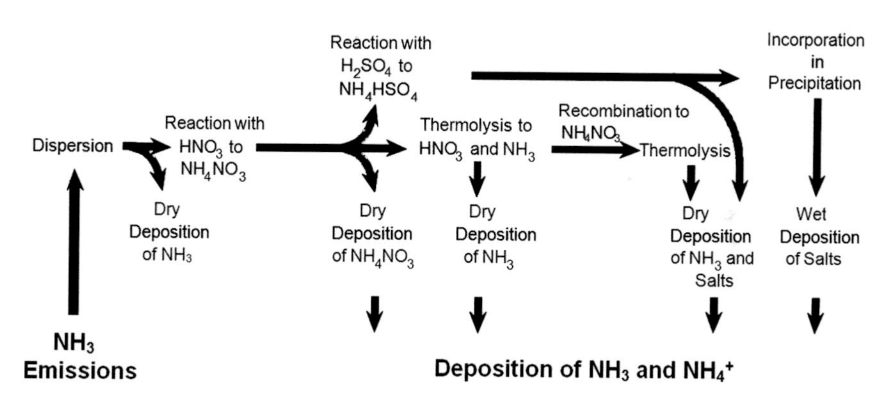

Fig. 2.9 Emission, dispersion, vertical and horizontal transport, chemical reactions, and deposition of ammonia and ammonium (*Amended, translated and adopted from* Dämmgen and Erisman 2002), where NH_3 is ammonia; HNO_3 is nitric acid; NH_4NO_3 is ammonium nitrate; NH_4HSO_4 is ammonium hydrogen sulfate; and H_2SO_4 is sulfuric acid

2.4 Hydrogen Sulfide

Hydrogen sulfide (H_2S), an aggressive trace gas, is generated from anaerobic breakdown of manure after some time in storage, where it is stored in the manure as gas bubbles (CIGR 1994). H_2S is highly toxic, poisonous, deadly, odorous (odor of rotten eggs/low concentrations contributed significantly to odor), colorless, and heavier than air, at low concentrations (<10 ppm). H_2S could cause dizziness, headaches, and irritation to the eyes and the respiratory tract. In addition to causing adverse effects to human and animal health, H_2S might be oxidized in the air forming sulfuric acid (H_2SO_4) resulting in acid rain that could cause ecological damage. H_2S concentration of 0.1 % can cause unconsciousness and death through respiratory paralysis unless artificial respiration is applied immediately. H_2S deadens the olfactory nerves (the sense of smell); therefore, if the smell of rotten eggs appeared to have disappeared, this did not indicate that the area was not still contaminated with this highly poisonous gas (CIGR 1984, 1994, 1999). Manure tank agitation is then followed by H_2S emission, and consequently, possible death occurs. Therefore, after manure tank agitation the area must be evacuated and the team members must leave the area.

2.5 Methane

There are two sources of methane production: (1) anaerobic decomposition of manure, and (2) enteric fermentation of fodder by anaerobic bacteria in of rumen where CH_4 is released by eructation. A cow's rumen produces 37 L of CH_4 per kg dry matter of feed intake. A cow digests 17 kg dry matter per day which release 0.5 m^3 CH_4 day^{-1} (CIGR 1994). Concentrations between 5000 to 15,000 ppm are explosive; several explosions have occurred due to ignition of methane-rich air in poorly ventilated livestock buildings. Figure 2.10 shows the anaerobic decomposition of organic matter (e.g., manure).

2.6 Carbon Dioxide

The CIGR report (1994) stated that the total CO_2 production is a sum of the following three components: animal respiration, rapid breakdown of urea in urine, and anaerobic decomposition of dry matter in the slurry. Over 96 % of the total CO_2

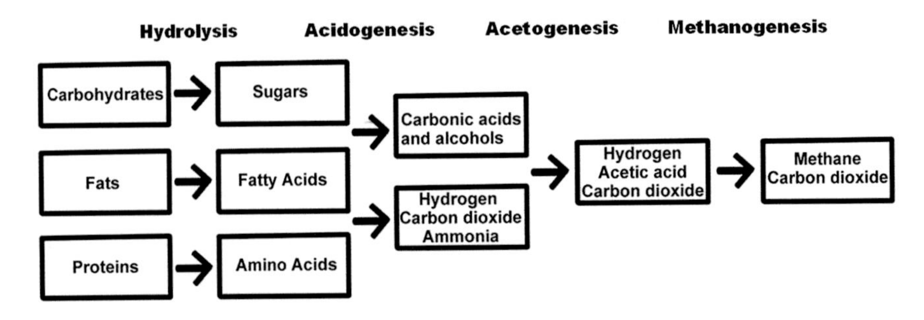

Fig. 2.10 Anaerobic decomposition of organic matter

production is from animals' respiration. Under normal conditions in livestock buildings CO_2 concentration is between 500 and 3000 ppm. There is no health risk for animals and humans at this level. The threshold limit is set to 3000 ppm.

2.7 Carbon Monoxide

Carbon monoxide is produced by incomplete combustion of fuels inside livestock buildings. For instance, when tractors operate some machines to accomplish an operation-like feed distribution and manure management. The threshold limit for CO is 10 ppm. Carbon monoxide can cause death in adult pigs at concentrations around 4000 ppm and in broilers at 2000 ppm (CIGR 1994).

2.8 Odors

The aerobic and anaerobic breakdown of organic substances (manure, feed left-overs) results in over 300 odorous components, whose mixture gives the smell impression in addition to the smells of animals and feed. The compounds of odors are produced from manure inside livestock buildings. Different gases are produced as livestock manure is degraded by microorganisms as previously maintained. Under aerobic conditions, CO_2 is the principal gas produced. Under anaerobic conditions, the primary gases are CH_4 and CO_2. About 60–70 % of the gas generated in an anaerobic lagoon is methane, and about 30 % is carbon dioxide. However, trace amounts of more than 40 other compounds had been identified in the air exposed to degrading animal manure. Some of these included mercaptans (this family of compounds included the odor generated by skunks), aromatics, sulfides, and various esters, carbonyls, and amines (CIGR 1999). Furthermore, odorous compounds in swine manure were ranged between 30 compounds that were the likely contributors of the odor nuisance and 168 compounds which had been identified by previous researches. The gases of most interest and concern in manure management are CH_4, CO_2, NH_3, and H_2S.

Manure handling and storage facilities can be a source of malodors in dairy operations. Offensive odor is partly the result of incomplete anaerobic decomposition of stored manure. Zhu and Jacobson (1999) found that the most important genera for odor production were *Eubacterium* and *Clostridium*. Studies have identified 35–73 volatile compounds in dairy manure (Filipy et al. 2006; Rabaud et al. 2003; Sunesson et al. 2001) with the most important odorous manure components found to be the volatile fatty acids (VFA), *p*-cresol, indole, skatole, along with hydrogen sulfide (H_2S) and ammonia (NH_3) by virtue of either their high concentrations or low odor thresholds (O'Neil and Phillips 1992). Wright et al. (2004) identified *p*-cresol, *p*-ethyl phenol, and isovaleric acid as the most persistent and biggest contributors to odor downwind of the source. Miller and Varel (2001) noted that ethanol, acetate, propionate, butyrate, lactate, and hydrogen were the major fermentation products of stored cattle manure. Due to far-reaching environmental and socioeconomic concerns, efforts to reduce odor, NH_3, H_2S, and greenhouse gas emissions from animal agriculture are essential (Wheeler et al. 2011b, c).

2.9 Dust and Aerosols

Aerosols can be defined as solid or liquid particles which remain suspended in the air for longer periods because of their minute dimensions of between 10^{-4} and 10^2 µm. The aerosols can combine chemically with gases emitted into the air and these new compounds are inhaled by living organisms. Airborne particulates can include both solid and liquid particles. Viable particles are living microorganisms or any solid or liquid particles which have living microorganisms associated with them. Dusts can be defined as dispersed particles of solid matter in gases which arise during mechanical processes or have been stirred up. Dust may cover a wide range of sizes and shapes, and can be airborne or settled (Hartung and Saleh 2007).

Generally, dust can be considered as one of the most important sources for air contamination in livestock buildings, where it may be generated from forages (ingredients, form, water, and fat contents), fur of animals (species, genotype, age, and number), bedding materials as litter (type, amount, and water content), dried manure, feathers/fur, dander (hair and skin cells), molds, pollen, grains, grain mites, insect parts, mineral ash, gram-negative bacteria, endotoxin, microbial proteases, ammonia adsorbed particles, infectious agents, and building materials (Robert 2001).

Dust formation on surfaces occurs by the effect of several forces, e.g., drying, chewing crushing cleaning, management (bedding, feeding, manure handling, etc.), and sedimentation. Further forces as animal activity, human activity, and airflow rates generate airborne dust in livestock building. The influencing factors are animal weight, animal density, housing system, ventilation system, daytime, and season (Aarnink and Ellen 2007). Affected by the ventilation, dust is emitted outside of the livestock building and this forms the dust emissions. Dust carries some pathogens, bacteria, and microbes. Additionally, some gases as NH_3 are adsorbed on the surface of the particulate matter (PM) of the dust. Figure 2.11 shows dust sources with attributes, processes, and forces that influence dust formation and dust emission from animal houses.

Airborne particulates can include both solid and liquid particles. Viable particles are living microorganisms or any solid or liquid particles which have living microorganisms associated with them. Dusts are dispersed particles of solid matter in gases which arise during mechanical processes or have been stirred up. Dust may cover a wide range of sizes and can be airborne or settled. Chemical properties of dust particles must be analyzed according to their chemical compositions which are divided into inorganic and organic (viable and nonviable) components.

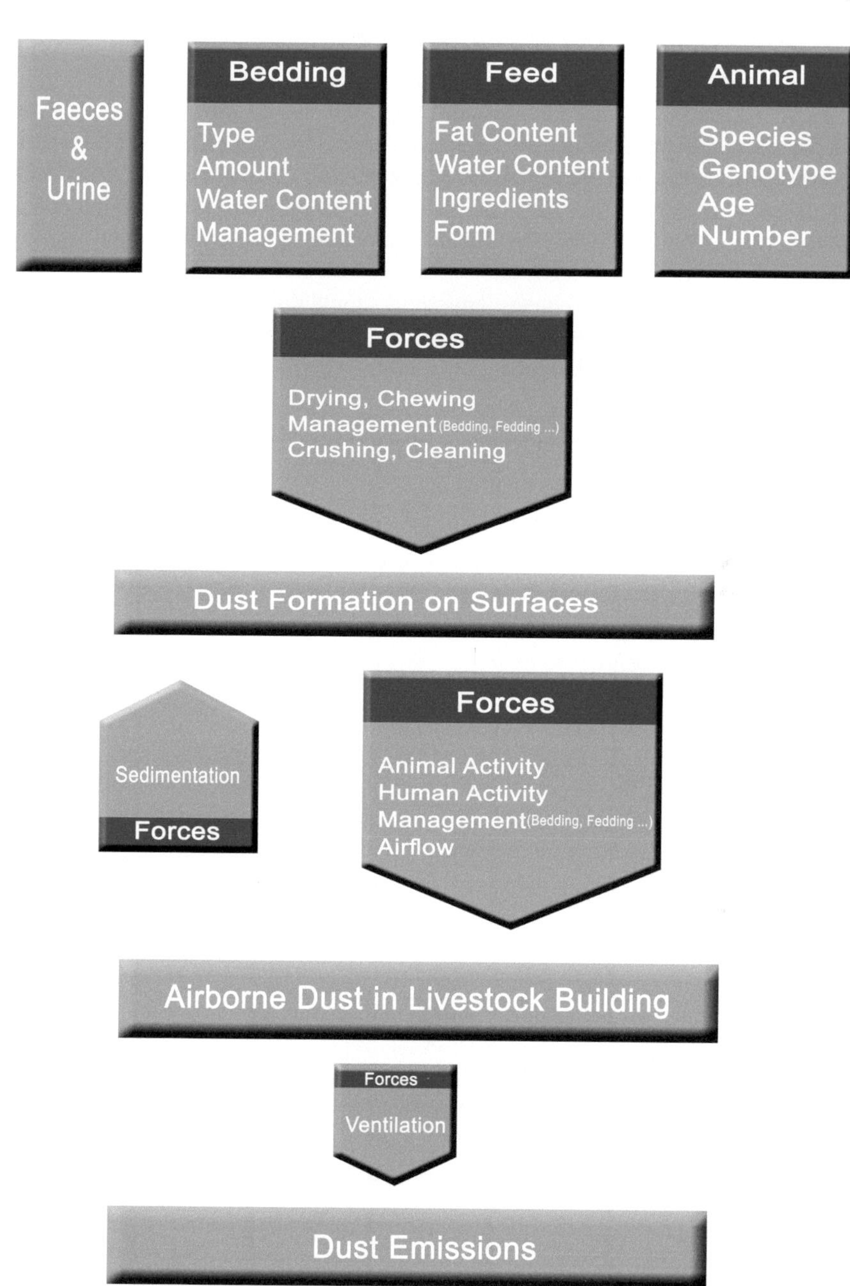

Fig. 2.11 Development, release, and spreading of dust inside the barn (*Amended, translated, redrawn and adopted from* Aarnink and Ellen 2007)

The chemical composition of dust from different sources shows that the airborne and the settled dust have nearly the same concentrations of Dry Matter (DM), ash, N, P, K, Cl, and Na. The dust particles are subjected to a variety of physical processes according to their density, size, and shape (Fig. 2.12). The most important physical effects said are sedimentation, agglomeration (particles collide due to the turbulence and adhere to each other forming agglomerates), aerodynamics, adsorption, and resuspension. The dust is characterized by sedimentation experimentation and microscopic analysis.

The dust contained in the exhaust emissions should not exceed a 20 mg/m^3 mass concentration or 0.20 kg/h of emission mass flow according to the maximum acceptable concentration (MAC) list (DFG 2006). The respirable dust (<5 μm) may

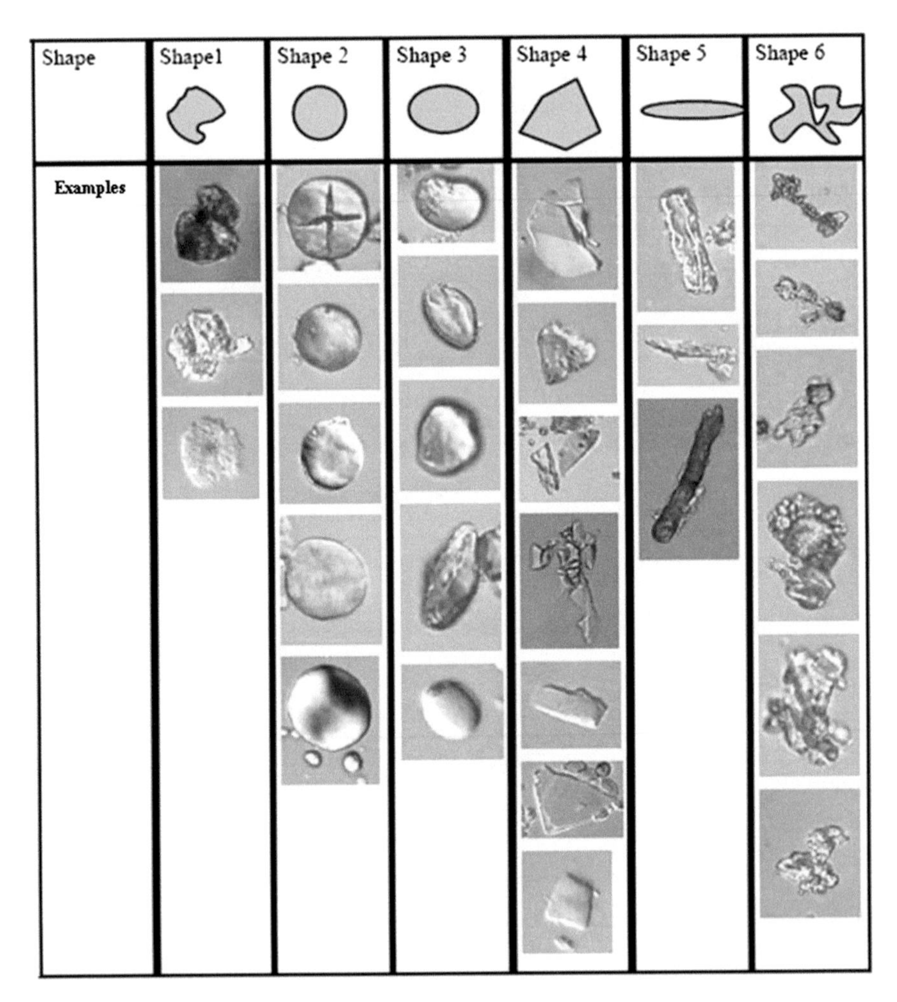

Fig. 2.12 Classification of particle structure after shapes (*adopted from* Mostafa 2008 *after* Nannen 2005)

not exceed a concentration of 4 mg/m^3 and for the alveolar dust (<1.1 μm) the limit value is 1.5 mg/m^3 (DFG 2006). Pedersen et al. (2000) showed that the limit recommendations for humans under Danish conditions are 2.4 mg/m^3 of total dust, 0.23 mg/m^3 of respirable dust with a total of 800 EU/m^3 (EU: Endotoxin Unit), and 7 ppm of ammonia. Endotoxins are toxins, poisonous substances produced by living cells or organisms, associated with certain bacteria.

An "endotoxin" is a toxin that, unlike an "exotoxin," is not secreted in soluble form by live bacteria, but is a structural component in the bacteria which is released mainly when bacteria are lysed. Lysis refers to the breaking down of a cell, often by viral, enzymatic, or osmotic mechanisms that compromise its integrity. Together with the dust particles microorganisms can be transported into the respiratory system causing infections. Endotoxins can trigger allergic reactions in the airways of susceptible humans even in low concentrations. PM carries odor, NH_3, endotoxin, bacteria, and fungi.

The hazard caused by aerosols, suspension of fine solid particles, or liquid droplets in a gas depends on their chemical composition as well as where they deposit within human respiratory system. Aerosols are solid or liquid particles which remain suspended in the air for longer periods because of their minute dimensions of between 10^{-4} and approximately 10^2 μm. The aerosols can combine chemically with gases emitted into the air and these new compounds are inhaled by living organisms or can settle on them. The mixture "air with dust particles" is considered as Newtonian fluid, where the flow of this mixture is treated as "single phase flow" in fluid mechanics. The Brownian motion rules the suspended aerosol particles in air, where the aerosols undergo irregular random motion due to bombardment by surrounding fluid molecules.

Pedersen et al. (2000) classified the dust into the following:

1. Total dust: the fraction containing particles below 20 μm in aerodynamic diameter collected by the use of 38 mm filter cassettes with 5 mm downward inlets.
2. Respirable dust: the fraction collected using a cyclone pre-separator (50 % cut-off effectiveness value of 5 μm).
3. Inhalable dust: the diameter of these dust particles is slightly larger than 20 μm. The inhalable concentration will be about 25 % higher than the "total dust" concentration, but it depends on the particle size distribution.

The airborne inhalable and respirable fractions are overall higher in pig and poultry buildings than in cattle houses. Dust concentrations and emissions are affected significantly by several things such as housing type, the season of year, and day/night time. The inhalable and respirable dust concentrations in the poultry buildings are 3.60 and 0.45 mg/m^3, respectively. The dust emission rates on a 500 kg AU are between 2118 and 248 mg/h for inhalable and respirable, respectively (Takai et al. 1998).

The particulate matters (PM) are categorized as PM_{10}, PM_5, $PM_{2.5}$, and PM_1. PM_{10} is particulate matter smaller than 10 μm aerodynamic equivalent diameter. Similarly, PM_5, $PM_{2.5}$, and PM_1 are particulate matters smaller than 5, 2.5, and

1 μm aerodynamic equivalent diameter, respectively. On the other hand, the total suspended particles (TSP) are tiny particles of aerosols or particulates at high concentrations in the air and could raise air pollution concerns. TSPs range in size from 0.001 to 500 μm.

There are several parameters that affect the dust formation and emission. The housing system and design affects indoor dust concentration and emission rate. For instance, the air in floor housing systems for laying hens may be more polluted than in traditional cage systems. The year seasons have significant effect on dust concentrations and emission rates where several studies showed that the dust concentrations are the highest in summer compared to the other seasons. On the other hand, the mean inhalable dust emission rates in winter and summer were estimated to be 1590 and 2388 mg/h for 500 kg live weight basis, respectively (Takai et al. 1998). The diurnal change and animal activity have significant effect on indoor dust concentration and emission rate, where Hessel and Van den Weghe (2007) found that the dust concentrations are twice as high during the light period (5542 μg/m^3) compared to the dark period (2598 μg/m^3). Indoor dust concentration is directly proportional to animal activity which is higher during lights-on. The ventilation rate greatly affects the indoor dust concentration and emission rate, where there is a high variation in the pattern of spatial dust distribution in mechanically ventilated pig buildings. Thus, the ventilation systems have direct effects on the spatial dust concentration, whereas the increase of the ventilation rate will not necessarily reduce the overall dust level effectively because the dust production rate will increase with increasing ventilation. The dust concentration can be measured using the following methods (Gustafsson 1997; Lim et al. 2003; Mölter and Schmidt 2007):

1. Gravimetric measurements of the amount of total dust (mg/m^3) with 37 mm diameter millipore filters at a flow rate of 1.9 L/min.
2. Counting the number of different sized particles with a Rion optical particle counter.
3. Weighing the settled dust on 0.230 m^2 settling plates.
4. Tapered element oscillating microbalance (TEOM).
5. Optical aerosol spectrometers (OAS).

Chapter 3
Emissions Abatement Techniques

Liquid manure storage facilities are sources of gaseous emissions of NH_3 and greenhouse gases especially CH_4 and N_2O. Methane is the most predominant greenhouse gas emission from liquid manure storage facilities (Samer et al. 2014b; Berg et al. 2006a). Therefore, several studies have investigated different mitigation strategies for reducing greenhouse gases and ammonia emissions. Emissions occur at all stages of manure management: from buildings housing livestock; during manure storage; following manure application to land; and from urine deposited by livestock on pastures during grazing. Ammoniacal nitrogen (total ammoniacal nitrogen, TAN) in livestock excreta is the main source of NH_3. At each stage of manure management TAN may be lost, mainly as NH_3, and the remainder passed to the next stage. Hence, measures to reduce NH_3 emissions at the various stages of manure management are interdependent, and the accumulative reduction achieved by combinations of measures is not simply summated. This TAN-flow concept enables rapid and easy estimation of the consequences of NH_3 abatement at one stage of manure management (upstream) on NH_3 emissions at later stages (downstream), and gives unbiased assessment of the most cost-effective measures. Ammonia can be converted into nitrous oxide at any stage of manure management.

The EMEP/EEA air pollutant emission inventory guidebook provides greenhouse emission abatement measures for animal husbandry and manure management in the form of best available technique (BAT) for each case, type of manure, land use, limits of applicability, emission reduction (%), and availability for different farms (EMEP/EEA 2009). On the other hand, the Executive Body for the Convention on Long-Range Transboundary Air Pollution (2007a, b) provided a guidance document on control techniques for preventing and abating emissions of ammonia. Moreover, the European Commission (2003) provided an Integrated Pollution Prevention and Control (IPPC) reporting on the BAT for intensive rearing of poultry and pigs.

Best available technique (BAT) has the following characteristics: (1) most effective protection of the environment as a whole (Best); (2) possible and viable implementation in relevant sector, taking into consideration the costs and

M. Samer, *Abatement Techniques for Reducing Emissions from Livestock Buildings*, SpringerBriefs in Environmental Science,
DOI 10.1007/978-3-319-28838-3_3

advantages (Available); and (3) design, construction, maintenance, and operation (Technique). The best techniques should provide/consider the following: (a) low emissions of pollutants (NH_3, CO_2, CH_4, N_2O, odor, aerosols, noise), (b) soils and water conservation (leakage control or no leakage), (c) efficient use of energy (for ventilation, heating, etc.), (d) efficient use of raw materials (feed, water, bedding materials, etc.), (e) animal welfare, (f) amount and quality of manure, (g) efficient use of waste, (h) possibility of technical application, (i) and feasibility.

Previous studies have evaluated different treatments and additives for reducing gaseous emissions from manure and slurry in laboratory using glass jars or plexiglass tanks and a multi-gas monitor (Samer et al. 2014b; Wheeler et al. 2011a, b; Reinhardt-Hanisch 2008; Berg et al. 2006a). On the other hand, reducing the emitting surface area, optimizing ventilation systems, implementation of cooling systems (hot season), and reducing manure retention time in building are the best available emissions abatement techniques for livestock buildings. The mitigation principles can be summarized as follows:

1. Influencing or interrupting the formation of pollutants:
   - Inhibiting the urease activity, where this will lead to reduce NH_3 formation.
   - Avoiding suboptimal conditions for the conversion of N-bonds, where this will result in N_2O reduction.
2. Feeding:
   - Reducing extensive cattle feeding, where this will result in CH_4 reduction.
   - Avoiding N-surpluses which cannot be metabolized and will be excreted.
   - Using adjusted rations to species and production stage.
   - Using feed additives (e.g., amino acids).
3. Housing system/storage/application:
   - Minimizing dirty (emitting) surfaces.
   - Covering manure storages.
   - Lowering the temperature and the pH value of manure.
   - Fast and frequent manure removal.
   - Fast incorporation into soils after application.
4. Ventilation:
   - Reducing air velocity and temperature near emitting surfaces.
   - Optimizing the balance between airflow rate, necessary air exchange rate, temperature needs, and animal welfare requirements.

The implementation of an emission abatement technique for reducing a particular gas may change the concentrations of other pollutants (e.g., dust, CO_2, H_2S, etc.), may have an effect on welfare and productivity of the animals, should be feasible and have low investment and operating costs, and should provide a safe environment for the farmers (CIGR 1994).

3.1 Livestock Buildings and Manure Management

Manure can be stored inside the livestock building in the form of either liquid manure stored in manure channel or solid manure as litter which consists of feces, urine, and bedding material (straw, sawdust, or wood shavings). Manure storage outside of the building can be as either liquid manure stored in lagoons or in above-ground tanks, or solid manure stored in field heaps or heaps in yards (Fig. 3.1).

According to CIGR (1994), the release of NH_3 within a livestock building can be minimized by reducing the evaporating surfaces, minimizing airflow rates above surfaces, lowering the temperature to be below 10 °C, lowering the pH of manure to be below 6, shortening the storage period inside the building, frequent manure removal from the building, and treating the manure aerobically for urine and feces separation or anaerobically for biogas production. The type of slats, animal behavior, the ventilation system, and the manure handling system affect one or more of the above-mentioned influencing factors. Generally, the larger the percentage of slotted area, the less the surface is available for evaporation and the lower the NH_3 volatilization will be. In case of using mechanical ventilation systems, the extraction of air beneath the slats reduces NH_3 concentrations inside the livestock building. Higher air velocity and lower NH_3 concentrations directly above the manure surface will, however, increase the release of NH_3. Therefore, low air velocity with stable airflow profiles inside the building will minimize NH_3 mass transfer from manure pits in the building. The applicability of the above-mentioned emissions abatement techniques depends strongly on the housing system and the building design. In the case cattle buildings, the following emissions abatement techniques can be implemented to reduce NH_3 volatilization: slopping floor with manure scraper, slopping floor with manure scraper under the slats, implementation of additives in manure storages, and reducing contact surface between manure and air by covering the manure storages. In case of poultry buildings, frequent removal

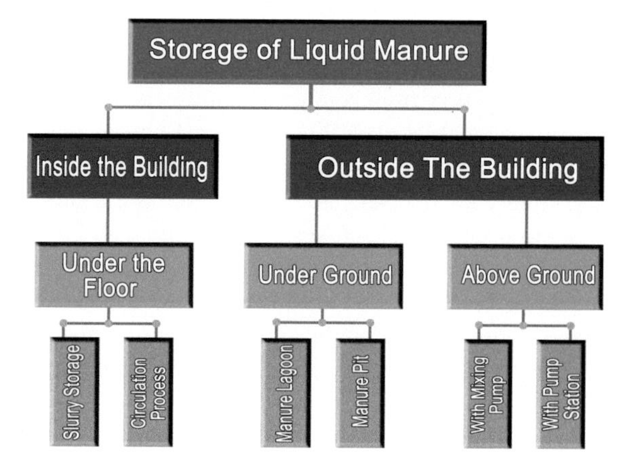

Fig. 3.1 Manure storage facilities

of manure would be the best emissions abatement technique. Dekker et al. (2011) stated that adjusting the housing system and the manure management has the potential to decrease the emissions. For instance, using organic laying hen husbandry in aviary systems instead of single-tiered systems has the potential to reduce emissions of NH_3, N_2O, and CH_4; further reductions might be realized by changes in litter management. UNECE (2007) stated that emission reductions can be achieved in poultry housing by drying manure and litter to a point where NH_3 is no longer formed by hydrolysis of uric acid.

Daily flushing of slurry from cattle houses would reduce total annual CH_4 and N_2O emissions by 35 % CO_2 equivalent, and that cooling of pig slurry in-house would reduce total annual CH_4 and N_2O emissions by 21 % CO_2 equivalent (Sommer et al. 2004). Methane emissions can be significantly reduced by complete slurry removal between the fattening periods and subsequent cleaning of the slurry pits in pig housing. Additionally, the release of methane from indoor slurry storage can be influenced by availability of oxygen and volatile solids, pH value, substrate temperature, retention time, and presence of inhibiting compounds. These factors should be further investigated to develop emissions abatement techniques. Special considerations should be given to avoid increasing specific gas emissions while abating another one (Haeussermann et al. 2006). For instance, abating agricultural emissions of NH_3 may cause the release of N_2O from this sector up to 15 % higher than in the case of no NH_3 control (Brink et al. 2001). Increased knowledge of the factors that affect emissions from livestock barns may lead to a better understanding of daily (between different days) and diurnal (within a specific day) variations in emissions, an improvement of mitigation methods, and a refinement of emission models. Animal activity, animal weight, indoor air temperature, and relative humidity have influence on carbon dioxide, methane, and ammonia emissions. Emission variations emphasized the need for measurements during different times within the day and during the growing period in order to obtain reliable data for assessing abatement techniques (Ngwabie et al. 2011).

In order to mitigate N_2O emissions from manure stored in animal houses, modifying the feeding strategy and adopting slurry-based system compared to a straw or deep litter-based system are the best mitigation methods. On the other hand, in order to mitigate CH_4 emissions from manure stored in animal houses, the following are the best mitigation methods: modifying the feeding strategy, frequent removal of slurry from beneath the house, and cooling slurry, e.g., below the slatted floor (Chadwick et al. 2011). In animal houses that do not use bedding materials, the slurry, feces, and urine remain in a predominantly anaerobic state with little opportunity for the NH_4^+ to be nitrified. As a result, little or no N_2O emissions are likely to occur from such systems (Zhang et al. 2005). Reducing the production of H_2S is possible by reducing the indoor temperature which will result in reducing the anaerobic breakdown of the manure stored inside the building. The indoor CO_2 concentration can be reduced by increasing the ventilation rate (CIGR 1994). However, this has an adverse effect where the emission rates of other noxious gases will increase.

Mathematical models and computer programs were developed to be implemented in constructing manure tanks and manure handling systems (Samer 2011a, 2008a) as well as biogas plants (Samer 2010), whereas the location of such systems in the farm vicinity was specified to be downwind to avoid gas transmission to the different farm facilities with a specified minimum distance between the farm and any adjacent residential communities, roads, and ecosystems (Samer 2008b). On the other hand, manure pits for temporary manure storage in livestock buildings form another effective source of gaseous emissions as shown in the different emissions inventories (Samer 2013a). Heat stress in dairy cows is one of the leading causes of decreased production and fertility. Increasing air velocity, using ceiling fans, to enhance convective heat transfer and accordingly body heat dissipation is highly required. However, this has negative effects such as increasing emission mass flux of the harmful gases. Airflow profiles affect the gas emission rates which increase with the increasing volumetric airflow rates and air velocities, where free air streams allow more gas release through convection mass transfer. Furthermore, the ceiling fans indirectly increase the ventilation rates which ultimately results in increasing the gaseous emissions. Therefore, a balance must be achieved among the different contradictions: air velocity optimization, heat stress alleviation, air distribution enhancement, and gaseous emissions reduction. On the other hand, gaseous emissions increase with increasing temperatures (Samer 2011b, 2012a; Samer et al. 2011a). The implementation of proper waste management which is safe to the surroundings fulfills the green building specifications (Samer 2013b).

A survey should be accomplished for the farms in a specific. This allows the effects of the variability of farm and manure management parameters among farms on GHGs and NH_3 emissions to be fully taken into account. Estimating the emission factors per animal for several livestock categories and different farm classes can be used to develop emissions inventory and to upscale available national inventory (Reidy et al. 2008a). The stratified sampling and the individual farm calculations allow the comparison of emissions from specific regions and altitudes and the study of the variability among farms. This approach permits a more detailed analysis of the regional distribution of GHGs and NH_3 emissions as well as a more robust and standardized monitoring of the future development of emissions. The emissions inventory can be then analyzed and implemented to develop effective GHG and NH_3 mitigation strategies focusing on the largest emissions sources.

Uncertainties of estimated emission factors (EF) should be assessed in order to update the annual CH_4 and N_2O emissions. Additionally, emissions from manure management have the largest uncertainty, due to the high natural variability of manure. The more the animal accurate data are available, the lowest the uncertainty is expected. This is the case in the intensified production systems (Merino et al. 2011). Several flow models were used to calculate GHGs and NH_3 emissions from litter-based systems and slurry-based systems. The variability of emissions found in practice is likely to be much greater for straw-based systems than for slurry systems. The differences in estimates of NH_3 emissions decreased as estimates of immobilization and other N losses increased. Since immobilization and denitrification depend also on the C:N ratio in manure, there would be advantages to include C flows in mass-flow models. This would also provide an integrated model for the estimation of emissions of methane, non-methane VOCs, and carbon dioxide. Estimation of these would also enable an estimate of mass loss, calculation of the N and TAN concentrations in litter-based manures, and further validation of model outputs (Reidy et al. 2008b, 2009).

It is crucial to determine the emission rates, fluxes, and factors before and after deploying an emissions abatement technique, where this should be considered in amending the present emissions inventory of the geographic area where the considered abatement technique has been deployed. Therefore, it is important to estimate the emissions of pollutants from livestock housing and manure management especially inside of the building. One key issue is to measure the concentrations of the pollutants in exhaust air as well as inlet air, at worker's level, animal's level, and close to manure storage inside the building. These measurements depend on season, daytime, and production stage. The characteristics of the measurements are duration, repetition, and measurements cycle (as often and as longer as possible). The indoor measurements should cover the concentrations of the different pollutants, ventilation rates, temperature, humidity, animal activity, and airflow profiles. The outdoor measurements should focus on the outdoor concentrations of pollutants and wind velocity, temperature, and humidity.

3.2 Additives

Liquid manure storage facilities, inside and outside of livestock buildings, are sources of gaseous emissions of NH_3 and greenhouse gases especially CH_4 and N_2O. Additives can reduce gaseous emissions from swine waste lagoons and pits. The additives have the potential to reduce methane emissions from anaerobic swine lagoons (Shah and Kolar 2012). Numerous types of amendments have been proposed to reduce odor and gas emissions. McCrory and Hobbs (2001) categorized commercial additives according to their modes of action: (1) digestive additives; (2) disinfecting additives; (3) oxidizing agents; (4) adsorbents; and (5) masking agents. Chemical pH adjustment additives are also used to manage off-gas emissions. Microbial digestive additives consist of selected microbial strains and/or enzymes that reduce production or enhance decomposition of odorous compounds in animal wastes. Despite the inconsistent, and sometimes ineffective, performance of commercial manure amendments, these products continue to be the most widely available and popular type of odor control.

Amendments can be practical and cost-effective for reducing NH_3 and GHG emissions from dairy manure. Amendment products that act as microbial digest, oxidizing agent, masking agent, or adsorbent significantly can reduce NH_3 by more than 10 %, whereas microbial digest/enzymes with nitrogen substrate appeared effective in reducing CH_4 fluxes. For both CH_4 and CO_2 fluxes, aging the manure slurry for 30 days can significantly reduce gas production. Some amendments reduced odor emission depending on the storage period (Wheeler et al. 2011b, c).

Low starch content of beef cattle feedlot manure limits VFAs production (Miller and Varel 2001).

The effectiveness of the Digest3+3© microbial additive for reducing odor and pollutant gas emission from a swine gestation-farrowing operation was evaluated, where the additive was used to treat the deep pits to be compared with other untreated pits. However, they found no significant differences in terms of odor, NH_3, and H_2S concentrations and emissions between treated and untreated units. Overall, the microbial treatment had very little effect in reducing odor, ammonia, and hydrogen sulfide emission (Rahman et al. 2011). Selected essential oils are being promoted as effective and safe antimicrobial or antiviral (disinfectant) agents that also act as masking agents in the control of odor. Essential oils are aromatic liquids extracted from plant material via expression, fermentation, or distillation methods (Burt 2004) and are known to have various modes of action.

Oxidizing agents transform odorous compounds into less-offensive gases by chemical oxidation. Strong oxidizing agents act as disinfectants through their ability to degrade enzymatic proteins and oxidize sulfides, mercaptans, and NH_3. In a study of ferric chloride ($FeCl_3$) on degradation of odorous compounds, Castillo-Gonzalez and Bruns (2005) reported a significant reduction of volatile fatty acids concentration (propionic butyric, isobutyric, valeric, and isovaleric) in swine manure between 2 and 6 days incubation at 25 °C. At concentrations of 480 and 240 mg/L of potassium permanganate ($KMnO_4$), Ritteret al. (1975) reported that the mixture was effective in controlling odors from dairy slurry. In a laboratory study, hydrogen peroxide (H_2O_2) caused a very significant reduction in *p*-cresol levels (Eniola et al. 2006). Govere et al. (2007) found complete removal of three phenolic odorants, without recurrence for 72 h, from swine waste via gas chromatograph analysis after the addition of a mixture of hydrogen peroxide and miniced horseradish, while odor intensity was cut into half as determined by a human odor panel. Generally, oxidizing agents are effective in reducing malodors, but only for a short period, due to the large quantities of reagents required for complete oxidation (Wheeler et al. 2011b, c).

Natural zeolite, clinoptilolite (an ammonium-selective zeolite), has been shown to enhance adsorption of volatile organic compounds and odor emitted from animal manure due to its high surface area. Cai et al. (2007) reported reduction >51 % for selected offensive odorants (i.e., acetic acid, butanoic acid, isovaleric acid, dimethyl trisulfide, dimethyl sulfone, phenol, indole, and skatole) in poultry manure with a 10 % zeolite topical application. However, it showed some ineffective performances. It is believed that the frequent poor performance of absorbents stems from selective odorant adsorption, leaving other noxious odors to escape.

An effective odor amendment must be inexpensive, efficient, and suitable to dairy farm management. Several of these amendments cause an increase in total solids in manure storage (i.e., adsorbents) or inhibit the natural degradation of solids by the indigenous microbial population (i.e., disinfecting or alkaline materials). Extra benefits of an effective odor amendment may offer farmers, in addition to odor and gas emission controls, improved manure handling properties, reduction in surface water pollution, and in some cases reduction in the levels of pathogenic bacteria with potential benefit in soil pH adjustment (Wheeler et al. 2011b, c).

Manipulating the balance between ammonia and ammonium by lowering the pH value of slurry is another measure to reduce emissions (Stevens et al. 1989; Oenema and Velthof 1993; Hendriks and Vrielink 1997; Kroodsma and Ogink 1997; Martinez et al. 1997; Beck and Burton 1998; Pedersen 2003). Ammonia and methane emissions can be controlled by pH value. Manipulating the pH value of slurry has an effect on the balance between ammonia and ammonium. The pH values of untreated slurries range between 7 and 8 usually. Lowering the pH reduces the gaseous emission. From former investigations, it is known that a slurry pH around 5.5 can reduce ammonia emission by 80–90 % (Al-Kanani et al. 1992; Berg et al. 2006a, b; Husted et al. 1991; Li et al. 2006; Pain et al. 1990; Stevens et al. 1989). A slurry pH below 4.5 nearly avoids ammonia emission (Hartung and Phillips 1994). The pH value influences the activities of microorganisms. Higher methane production occurs, when the pH value is between 6 and 7 (Lay et al. 1997). A slurry pH below 6 is necessary to reduce methane emission and below 5 impede methane formation (Berg et al. 2006a, b). Whereas the use of inorganic acids has several disadvantages, using organic acids is a promising possibility to reduce not only ammonia but also methane and nitrous oxide emissions (Berg and Hoernig 1997; Berg and Pazsiczki 2003; Berg 2003). Samer et al. (2014b) developed a hypothesis for reducing emissions from manure stored inside and outside of livestock buildings. The hypothesis is treating manure with acidic liquid biowastes (e.g., wastes of citrus and milk industries) where the organic acids in the liquid biowastes will reduce the pH of manure which consequently mitigates gaseous emissions. Eventually, this process is an integrated waste management of both manure and acidic liquid biowastes. Therefore, the objective of this research paper is to investigate the possibility of reducing gas emissions (CH_4, N_2O, and NH_3) from dairy manure by adding low-pH biowastes, e.g., whey, wastes of citrus juice industries, etc.

The effect of NI dicyandiamide (DCD) on transformations of N to nitrate (NO_3^-) was investigated and subsequent reduction to N_2O in a grazed pasture system receiving cow urine, where the DCD was able to decrease the nitrification rate (Giltrap et al. 2010). Based on this study, an issue can be raised: is the DCD able to be used as an inhibitor of N_2O emissions from floor inside a livestock building? A promising approach for reducing ammonia emissions from dairy farming is the use of urease inhibitors. The basic investigations on urease inhibitors afforded an important contribution to the expansion of knowledge in this area, and will lead on the other hand to develop new techniques in order to reduce the NH_3 emissions from livestock housing (Reinhardt-Hanisch 2008). The implementation of urease inhibitors is effective in reducing ammonia emissions from cattle and pig slurry (Hagenkamp-Korth et al. 2015).

3.3 Covering Manure Storages

Different materials for covering liquid manure storage facilities have been investigated and are in use for mitigating odor and ammonia emissions (Sommer et al. 1993; Williams 2003). These materials abate also methane and nitrous oxide emissions. Different materials for covering liquid manure storage facilities to reduce gaseous emissions were investigated on laboratory scale: perlite, lightweight expanded clay aggregate, and chopped straw—both individually and combined with lactic acid or saccharose, respectively (Berg et al. 2006a). Covering pig manure with pulverized lignite reduces NH_3 emissions by 70 % and odor emissions by 50 % with no mitigation of GHG (Berg and Samer 2010). Nitrous oxide is emitted at one-tenth the rate of methane. However, it can increase using common cover materials (straw and granules) which reduce ammonia emissions effectively. The higher N_2O emission rates occur when manure tank has a dry encrusted surface. Hence, the strongest encrustation delivers the highest N_2O emission fluxes. Adding water to the encrusted surface, simulating rainfall, could reduce nitrous oxide emission (Berg et al. 2006a).

3.4 Aerobic and Anaerobic Treatment

Manure can be treated and utilized as a biofertilizer (Fig. 3.2). Liquid manure treatment can be aeration, separation, or anaerobic digestion. Solid manure treatment can be active composting or anaerobic digestion. The anaerobic digestion of manure for producing biogas and biofertilizer is an effective emission abatement technique. The generally positive impacts of anaerobic and aerobic treatment on the reductions of methane and volatile organic compounds (VOCs) are confirmed. However, the effects of anaerobic and aerobic treatment varied over the time of storage, especially for VOCs. In order to achieve significant reductions in VOCs emission, the storage time of anaerobic digester or aerobic reactor effluent should be limited to no more than 84 days (Zhang et al. 2008). Several studies discussed the plan and design of biogas plants and household units (Samer 2010, 2012c). In order to comprehensively estimate the significance of biogas utilization on rural energy development and greenhouse gas emission reduction, Yu et al. (2008) analyzed all types of energy sources, including straw, fuelwood, coal, refined oil, electricity, liquefied petroleum gas (LPG), natural gas, and coal gas, which were substituted by biogas, based on the amount of consumption. Energy substitution and manure management working in combination, i.e., coupled issues of environment and energy, reduce the emissions of GHGs efficiently. By the employment of biogas digesters, the reduction of GHGs (CH_4, N_2O and CO_2) was estimated to be 49.7 % of CO_2 equivalents (CO_2−eq).

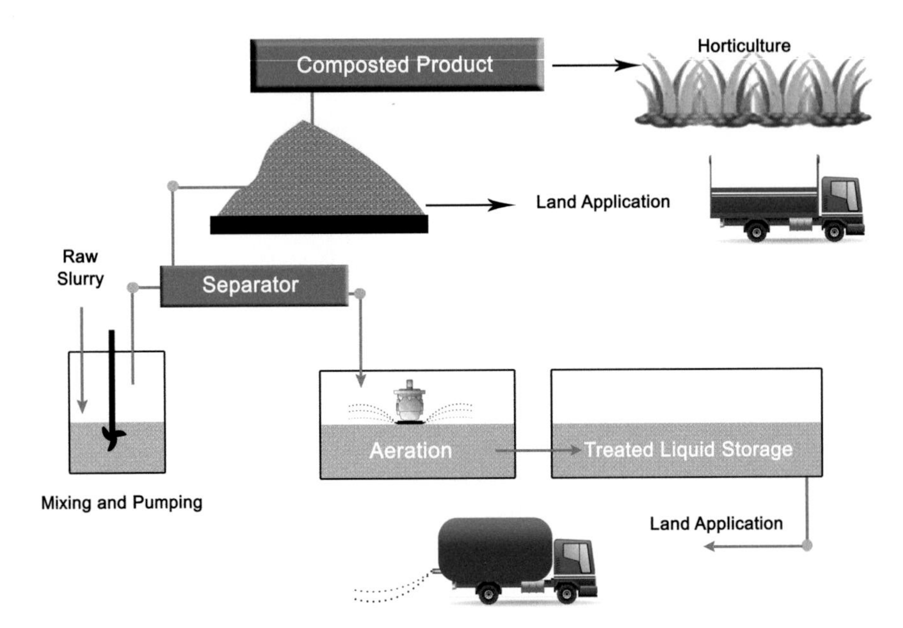

Fig. 3.2 Aerobic treatment and utilization of livestock manure (*Amended, translated, redrawn and adopted from* Burton et al. 2003)

3.5 Dietary Manipulation

The primary source of NH_3 is nitrogen (N) in the feed. The better the nitrogen utilized by an animal, the less the nitrogen will be excreted by the animal. It is recommended that the exact amounts of amino acids in the diet be provided to meet the nitrogen requirements of the animal. However, the animal requirements change with age and weight (CIGR 1994). Manipulation of livestock diets to affect manure production, composition, and odor is an effective emission abatement technique. Nitrogen excretion rates, which affect N_2O emissions from manure, are based on dry matter intake (DMI) through diet (Vergé et al. 2012). Therefore, diet manipulation to improve animal N utilization efficiency is one of the most effective measures to reduce livestock NH_3 emissions compared to housing and manure storage techniques (Carew 2010). Similarly, it is an effective measure to reduce N_2O emissions. On the other hand, the reliance of beef production on roughages makes enteric methane a major term in the GHG budget of beef cattle (Capper 2012; IPCC 2006). There is currently no effective mitigation for enteric methane emissions (Cottle et al. 2011). However, more grain in the cattle diet reduces the intensity of enteric methane emissions (Capper 2012). On the other hand, the primary source of ammonia is nitrogen in the feed. The better the nitrogen utilized by an animal, the less the nitrogen will be excreted by the animal. It is recommended that the exact amount of amino acids in the diet be provided to meet the nitrogen requirements of the animals. However, these requirements change with age and weight (CIGR 1994).

3.6 Dust Emissions Abatement Techniques

Molecules of the greenhouse gases (CH_4, N_2O, NO, and CO_2) and ammonia can be adsorbed by the particulate matters emitted from livestock housing and from the surface manure stored temporarily in livestock buildings as well as from the contaminated surfaces inside the livestock buildings. Therefore, dust emissions are another mean of greenhouse gases and ammonia emissions. Consequently, reducing dust emissions is essential. On the other hand, the EU emission norms for PM have come into effect recently, may be limited for continuation and/or expansion of intensive livestock operations in the near future, alongside existing ammonia and odor emission standards (Melse et al. 2009). Therefore, the implementation of dust abatement techniques is crucial.

Generally, the mitigation strategies are limitation of the particle release, bonding of particles, limitation of re-suspension, and filtering/cleaning of air. The abatement techniques are spraying of oil and/or water, adding water or oil to feed, feeding technique and frequency, optimization of air distribution, little straw use, and implementation of air filter. Furthermore, increasing the air quality can be achieved by purifying the re-circulating air inside the animal barn. There are dry and wet filter techniques. The dry filter achieves the highest dust reduction efficiency in comparison to the cyclone and wet filter system. The reduction efficiencies of the dry filter, under commercial scale barn measurement, are 55 and 72 % for indoor concentration and dust emission rate, respectively.

There are several means for reduction of dust in/from livestock houses, such as (1) dust suppression with spraying oil and/or water, (2) ionization, (3) cleaning the air by oxidization using chemical compounds called "oxidants," (4) use of windbreaks, and (5) implementation of dedusters, scrubbers, and filters as either dry or wet (Mostafa 2008; Mostafa and Buescher 2011; Mostafa 2012). The most perspective options to reduce dust emissions from animal houses are as follows (Aarnink and Ellen 2007):

1. Source problem solving: feed (using improved pellets, coating pellets, liquid feed), feces and urine (reducing pen fouling), and bedding refreshment.
2. Preventing dust formation by preventing manure drying and improving processes for making and transporting feed and straw.
3. Preventing dust to become airborne by reducing activity, improving feed distribution system, adding oil to animal feed, spraying oil and/or water, making big layer of bedding material, and optimizing pen design.

4. Preventing dust emission by using internal air cleaning systems (air filter and electrostatic filter), implementing external air cleaning systems (biofilter, bio-scrubber, air filter, electrostatic filter, water curtain, mist of water).

3.6.1 Spraying Oil and Water

Fogging (only water) is used for reducing indoor dust. By means of nozzles water can be sprayed in the barn. Spraying just before the herdsman enters the milked cows to the barn reduces the dust. Additionally, showering the passages and fittings will moisten the dust to a level where the dust cannot rise again (CIGR 1992). Showering water on floor surfaces in the walking alleys reduces the total dust concentration by 9 % and spraying salt solution of KCl in the air using nozzles reduces the total dust concentration by 41 % (Gustafsson 1997). Zhu et al. (2005) stated that dust concentration in swine gestation houses can be reduced by 75 % of average airborne dust concentration in the summer season through spraying/misting during the feeding time. Fogging by means of water has a short effect on the amount of dust in the indoor air due to the evaporation of water. A more permanent effect can be obtained using oil, because the evaporation of oil is slow (CIGR 1992).

Spraying a mixture of oil and water was proved to be a very effective method for reducing dust in livestock buildings at relatively low costs. The main effect of oil and water spraying is preventing dust on surfaces to become airborne. Using a quality design, dust reduction can reach 90 % (Pedersen et al. 2000; Aarnink and Ellen 2007). A variety of vegetable oils including canola, corn, sunflower, flax, soybean, and rapeseed oils along with mineral oils have been used to control dust from feed sources and building floors. Soybean oil reduced dust counts by 99 % following 1–2 % addition to dry feed (Pedersen et al. 2000; Ullman et al. 2004). On the other hand, spraying mixtures of oil and water in pig houses reduces dust concentration by 75–80 % (Gustafsson 1997).

3.6.2 Oxidizing Agents

An oxidant, oxidizing agent, can be defined as a chemical compound that readily transfers oxygen atoms or a substance that gains electrons in a chemical redox, i.e., reduction/oxidation reaction (Mostafa 2008, 2012). Cleaning the air can be achieved by oxidation using oxidants such as ozone, potassium permanganate, chlorine, and chlorine peroxide. Ullman et al. (2004) evaluated indoor ozone system for dust control effectiveness, where the total dust concentrations decreased by 60 % at the fan exhaust under maximum tunnel ventilation compared to a nearby building without ozone treatment.

3.6.3 Ionization Systems

When negative ions are discharged by ionization heads installed in a livestock barn, the dust particles will be ionized and attracted to earth-connected surfaces (CIGR 1992). Precisely, ionization is the physical process of converting an atom or molecule into an ion by adding or removing charged particles such as electrons or other ions. This process works slightly differently depending on whether an ion with a positive or a negative electric charge is being produced. A positively charged ion is produced when an electron bonded to an atom (or molecule) absorbs enough energy to escape from the electric potential barrier that originally confined it, thus breaking the bond and freeing it to move. The amount of energy required is called the ionization potential. A negatively charged ion is produced when a free electron collides with an atom and is subsequently caught inside the electric potential barrier releasing any excess energy (Mostafa 2008, 2012).

The reduction of dust concentration in animal buildings using an ionization system was investigated, where the ionization of air imparts a negative charge on dust particles that can then be attracted to collection plates or rods. Ionization reduces dust concentrations by about 78 %, with reductions ranging from approximately 68–92 % for six different ranges of ionization (Ullman et al. 2004). Electrostatic space charge systems (Fig. 3.3) were shown to remove up to 91 % of artificially generated dust and 52 % of dust generated by mature White Leghorn chickens in a caged layer room. An apparatus consisting of two negatively charged needles located 0.25 m above the floor and a positively charged aluminum collector plate (0.76 m high by 1.4 m long) located in front of the door, charged at 12 and

Fig. 3.3 Electrostatic space charge system (*Adopted from* Mitchell and Baumgartner 2007)

8 kV, respectively, was tested at a livestock facility. Ionization was approximately 6 times greater at dust removal than gravity alone.

An electrostatic space charge system was investigated by Mitchell et al. (2004) to demonstrate its effectiveness for implementation in a breeder/layer farm environment for reducing airborne dust. The system implemented ceiling fans to distribute negatively charged air throughout the room and to move negatively charged dust downward in the direction of the grounded litter where most of the dust would be captured. The dust concentration was reduced by an average of 61 % over a period of 23 weeks.

3.6.4 Aerodynamic Dedusters

Zhang et al. (2001) developed two aerodynamic uniflow dedusters (a cyclone-type particle separator and gas remover with airflow capacity 188 and 1880 L s^{-1}) with low pressure requirement and high particle separation efficiency. This development is based on fluid dynamics, particle mechanics, and sensitivity analysis. The small model deduster employs a set of turbine-type vane guides, an involute separation chamber, and a flow converging section to minimize turbulence and reduce the pressure loss. Dusty air is drawn from the air inlet passing through a set of vanes to establish a spiral flow pattern. The air then passes through the involutes' chamber and converges at the exit section above the dust bunker. Particles are collected in the dust bunker and clean air is exhausted through the blower. This device unlike the conventional cyclones can remove respirable particles at pressures of 50 Pa. The large model deduster contains three concentric dedusters. The outer cylinder of the smaller deduster serves as the inner cylinder of the bigger deduster. Thus, the total cross-sectional area is increased to allow air delivery and the volume of the unit is minimized. The fan speed can be varied via a frequency controller so that the performance at different airflow rates can be evaluated. An automatic dust flushing system was developed to periodically clean the dust in the dust bunker. The new design is aimed at reducing dust emissions for exhaust fans with large air flow rates. The dust mass concentration was measured at the inlet and the outlet of the deduster using filter collectors during 24 h periods. The results showed that the dust mass removal efficiency was 91 % at the 60 % power level. The dust reduction efficiency was 89 % at 100 % power level. Figure 3.4 shows aerodynamic dedusters. Air was drawn into the annular tunnel through a set of vane.

3.6.5 Bioscrubbers

Snell and Schwarz (2003) described the exhaust air cleaning system based on a bioscrubber. The exhaust air flows horizontally to the house gable and passes through the fans to enter the filter which is located outside the stable. In the

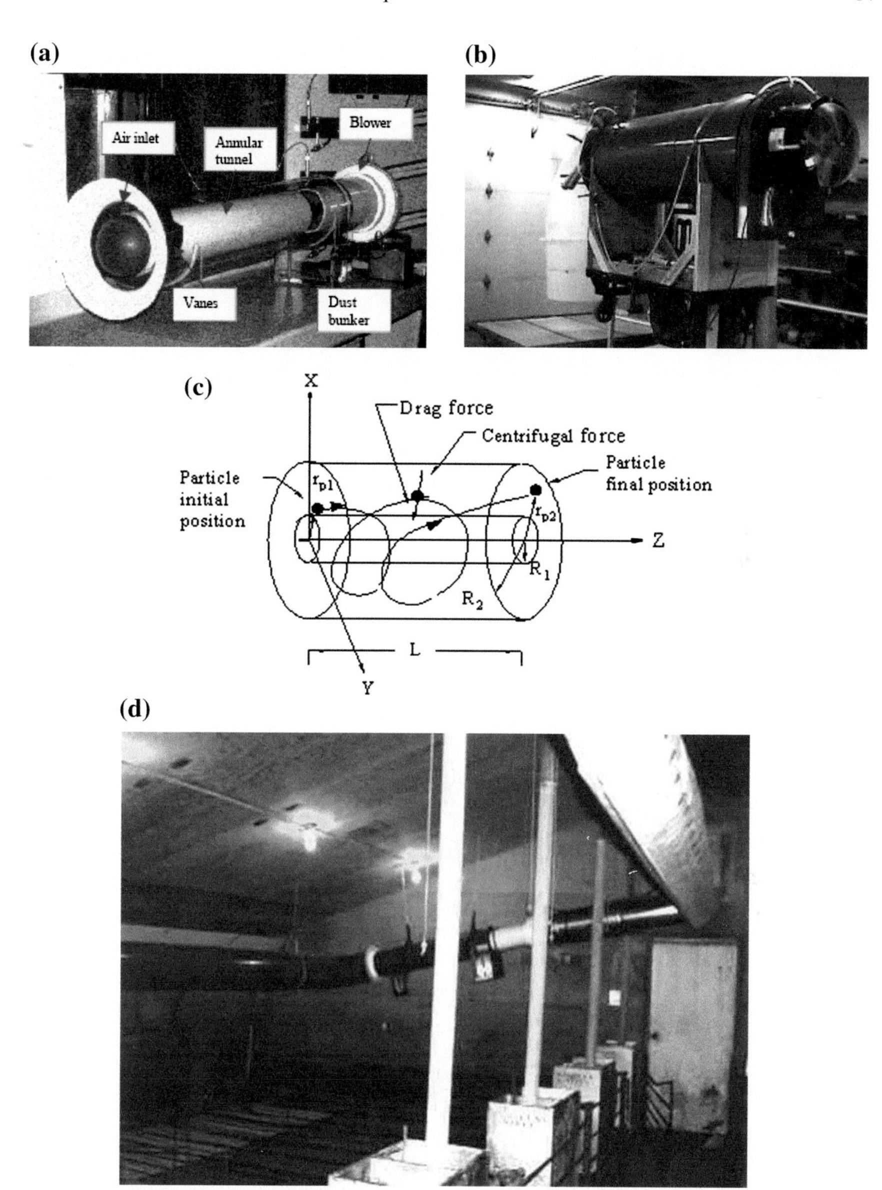

Fig. 3.4 Uniflow deduster where air is drawn into the annular tunnel through a set of vane. **a** First prototype (Zhang et al. 2001), **b** Large prototype (Zhang et al. 2001), **c** Forces affecting the particle trajectory inside the deduster (Zhang 2000), **d** Air cleaning recirculation system with dedusters (*Source* Illinois Odor and Nutrient Control Proving Center)

beginning the air is humidified and then flows into the first filter bank which consists of the so-called pads. In this stage the dust is washed out of the air and transported downward by the water and the air flows through the second filter bank.

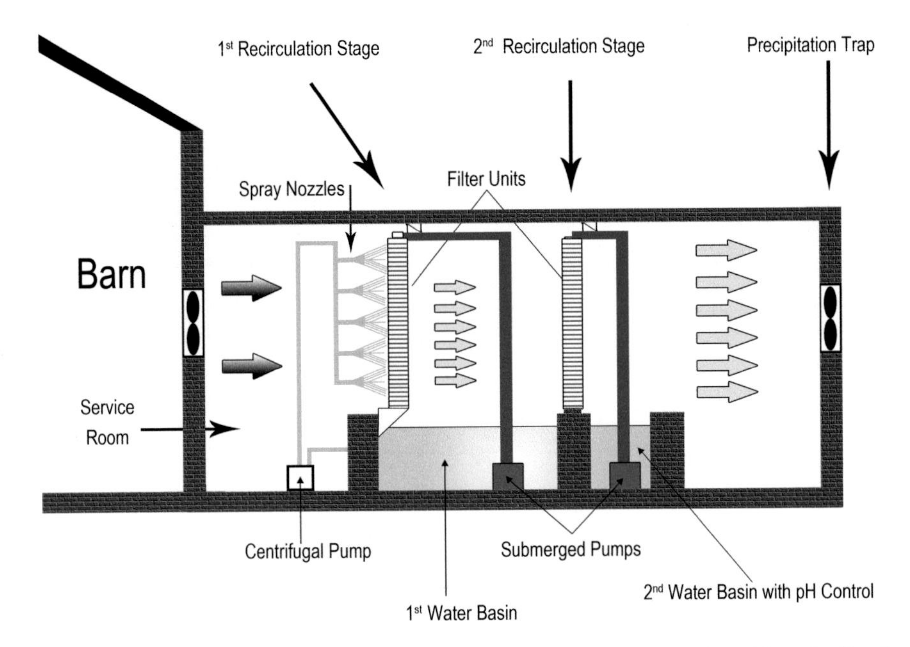

Fig. 3.5 Exhaust air cleaning system (*Amended, translated, redrawn and adopted from* Snell and Schwarz 2003)

In this filter the pH value of the water is regulated by acid to eliminate NH_3, fine dust, and odorous substances which cannot be washed out in the first filter bank. The water from both filters is collected and smoothed so the solid matter deposits on the ground of the basin and the water is then pumped up to flow over the pads again. The results of the dust concentration measurements show that more than 80 % of the airborne dust was removed by the filter. Figure 3.5 shows the components of exhaust air cleaning system.

3.6.6 Windbreak Trees and Walls

A windbreak or shelterbelt is a plantation usually made up of one or more rows of trees planted in such a manner as to provide shelter from the wind (Mostafa 2008, 2012). Windbreak walls placed at 3–6 m from the building deflected the airflow from the exhaust fans in the upward direction similar to other wind barriers, and therefore, it provides area for dust deposition. The vertical height at which the dust plume would flow over a downwind area under low wind velocity was increased by windbreak walls. Consequently, the dust levels in the downwind area from the windbreaks were minimized (Pedersen et al. 2000).

3.7 Biofiltration for Odor Control

There are a variety of approaches to removing odors and gaseous pollutants from effluent streams: absorption, adsorption (gas adsorption is used industrially for odor control), condensation, chemical and/or biochemical reaction, incineration, and selective diffusion through a membrane. One of the implemented techniques is the elimination of malodorous gas emissions (odorous compounds, hydrogen sulfide, and ammonia) and volatile organic compounds (VOCs) from a livestock building using a biofilter, which is also an attractive technique for the reduction of greenhouse gases. Chen and Hoff (2009) stated that biofilters can be used as an effective technology for reducing odor and VOCs emissions from animal facilities, where the reduction efficiency (RE) is up to 99 % for odor and up to 86 % for odorous VOCs.

The removal of methane from exhaust air of livestock buildings and manure storages (air from the headspace of liquid manure tanks) has a large potential for the reduction of greenhouse gas emissions from animal husbandry. Melse and Van Der Werf (2005) designed a biofilter with a filter bed consisting of a mixture of compost and perlite in a 40:60 (v/v) ratio which was inoculated with activated sludge that had shown a good methane oxidation rate as compared to pure cultures. Methane removal up to 85 % could be achieved. The methane removal (g m^{-3} h^{-1}) appeared to be proportional to the concentration (g m^{-3}). Relatively low methane concentrations and high air flows, as reported for the exhaust air of animal houses, would require very large biofilter sizes. Treatment of air from 1000 m^3 liquid manure storage with a methane concentration of 22 g m^{-3} would require a 20 m^3 biofilter for a desired emission reduction of 50 %.

The operating principle of a biofilter is that the contaminated air from the building is passed through a chamber which contains a moist packing medium. Provided moist conditions are maintained, naturally occurring aerobic bacteria populate the packing medium. As the air flows through the biofilter, the undesirable components are dissolved in the moisture on the packing medium where they are sequestrated, where the aerobic bacteria oxidize them, forming principally carbon dioxide, water, and mineral salts (Phillips et al. 1995). The water and humidity of the packed column inside the media of biofilters are very important for the process of absorption and to offer the microorganisms an optimal bioenvironment for growth (Jungbluth and Buescher 1996).

The biofilter main function is to bring microorganisms into contact with pollutants contained in an air stream. The box that makes up this biofilter contains a filter material, which is the breeding ground for the microorganisms. The microorganisms live in a thin layer of moisture, called the biofilm, which surrounds the particles that make up the filter media. During the biofiltration process, the polluted air stream is slowly pumped through the biofilter and the pollutants are absorbed into the filter media. The noxious gases diffuse into the biofilter media and are adsorbed onto the biofilm. This gives microorganisms the opportunity to degrade the pollutants and to produce energy and metabolic byproducts in the form of CO_2 and H_2O. The biofilter treats a contaminated air stream by biologically destroying the contaminants. The applicability of biofilters depends on the nature and the concentration of the organic constituent.

Biofiltration technology has been proven to be the most cost-effective method for treating ventilation exhaust air reducing emissions from livestock (Phillips et al. 1995; Nicolai and Janni 1999; Chen and Hoff 2009). Additionally, biofiltration has been proved to be effective for removing low concentrations of easily biodegradable constituents from air. Together with other advantages such as low cost and less maintenance requirement, biofiltration is a promising air pollution control (APC) for reducing odor emission. Several researches have concentrated on media choice, retention time determination, control of operating parameters, and reducing the capital costs. The biofilter can remove H_2S in air stream up to 120 ppm to below 1 ppm at airflow rate of 0.57 $m^3 s^{-1}$ (Yang and Tugna 1999).

Abd El-Bary (2003) elucidated the advantages and disadvantages of biofiltration process. The main advantage of using biofiltration over other more conventional control methods are lower capital costs, lower operating costs, low chemical usage, and no combustion source. Biofiltration units can be designed to physically fit into any industrial setting. A biofiltration unit can be designed as any shape, size, or as an open field with the piping and delivery system underground. In addition, biofilters can be designed with stacked beds to minimize space requirements and multiple units can be run in parallel. Biofiltration is versatile enough to treat odors, toxic compounds, and VOCs. The treatment efficiencies of these constituents are above 90 % for low concentrations of contaminants (<1000 ppm). Different media, microbes, and operating conditions can be used to tailor a biofilter system for many emission points. On the other hand, the capital costs of biofilter systems are comparable to those of alternate technologies such as air scrubbers and adsorption systems. Operating costs are minimal. There is no chemical or energy consumption in the process, and minimal instrumentation and monitoring are required. The major costs are filter bed media replacement every 3–5 years representing a depreciation of 10 % of total capital cost per year and electrical costs for operation of exhaust fans. On the other hand, the disadvantages of biofiltration lie in that biofiltration cannot successfully treat some organic compounds, which have low adsorption or degradation rates. This is especially true for chlorinated VOCs. Contaminant sources with high chemical emissions would require large biofilter units or open areas to install a biofiltration system. Sources with emissions that fluctuate severely can be detrimental to the biofilter microbial population and overall performance. Acclimatization periods for the microbial population may take weeks or even months, especially for VOCs treatment.

3.7.1 Biofilter Design

Biofilters can be classified as open or closed by configuration or as vertical or horizontal by gas flow direction (Chen and Hoff 2009). The common biofilter design is a large box next to the livestock building (Fig. 3.6). The most common

Fig. 3.6 Biofilters (*Source* Hartmann Filter GmbH & Co.KG.)

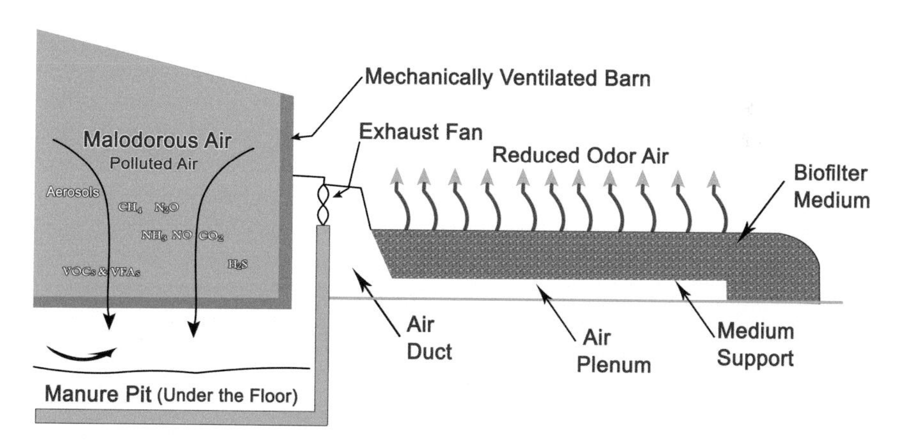

Fig. 3.7 Vertical gas flow open-bed biofilter (*Amended, translated, redrawn and adopted from* Schmidt et al. 2004)

types of biofilters are the vertical gas flow open-bed biofilter (Fig. 3.7) and the horizontal gas flow open-bed biofilter (Fig. 3.8). The vertical gas flow biofilter can be further divided into up-flow or down-flow, where the up-flow type is generally

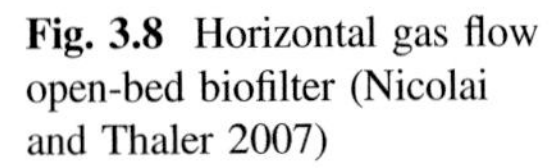

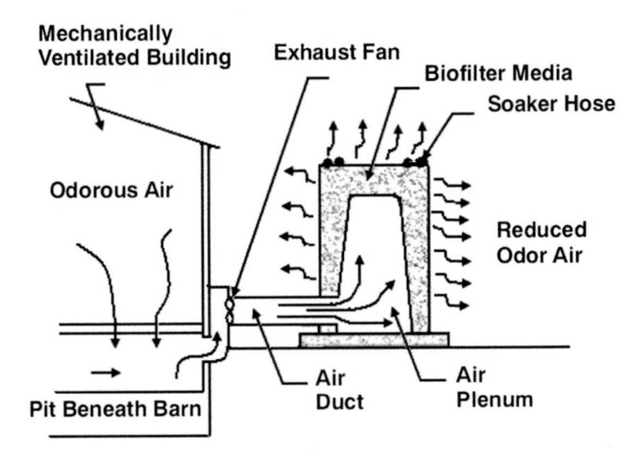

Fig. 3.8 Horizontal gas flow open-bed biofilter (Nicolai and Thaler 2007)

cheaper than down-flow in terms of construction costs (Nicolai and Lefers 2006). Therefore, up-flow open-bed biofilters are preferred for agricultural uses. The horizontal gas flow biofilters offer an option if enough surface area and space are not available. Chen and Hoff (2009) stated the factors concerned in design and operation of biofilters which are media property, empty bed residence time, media moisture measurement and control, microbial ecology, construction, and operation cost. The performance of a biofilter is evaluated in terms of odor and VOCs reduction efficiency (RE) and air pressure drop. The most important factors that affect the performance of a biofilter are packing media, media moisture content, and empty bed residence time. In order to install and operate a biofilter the following parameters should be considered: removal efficiency, air pressure drop, and construction/operation costs. The controlling factors are moisture content, temperature, pH of the medium, nutrients, contaminant load, oxygen content, airflow direction, dust, and grease, whereas the operation factors are (1) retention time (RT) which is the time length allowed to the contaminated air to be in contact with the biofilter media; (2) dynamic biofilter performance also called dynamic behavior of biofilter; (3) efficiency and performance of biofilters where the biofilter is most sensitive to interstitial velocity, biofilter height, specific surface area and first order biodegradation rate constant; and (4) biofilter clogging. The design parameters of biofilters are as follows:

1. Size and space: the space on-site is the greatest concern in designing biofilters. A small biofiltration unit can be designed to handle approximately 60 $m^3 h^{-1}$ in as little space as 2.33 m^2. Similarly, a biofiltration system, designed to treat large air volumes, requires space as large as a basketball court.
2. Biofilter depth: higher media depth has higher potential reduction efficiency with a maximum value. However, higher media depth results in higher pressure drop which is linearly related to media depth at a constant airflow rate. The media depth of 0.25–0.50 m has been recommended as optimal for agricultural biofilters (Chen and Hoff 2009).

3. Retention time (RT): the retention time represents the length of time the bacteria are in contact with the contaminated air stream and is equal to the ratio of void volume to volumetric flow rate. Therefore, the longer the retention times, the higher the efficiencies; however, the design must minimize retention time to allow the biofilter to accommodate higher airflow rates. For most biofilters, retention times range between 30 s and 1 min (Nicolai and Janni 1999).
4. Empty bed residence time (EBRT): EBRT is defined as the volume of biofilter media divided by airflow rate passing through the media. At a typical 5 s empty bed residence time and 55 % media moisture content, a mixture of compost and wood chips can achieve average reduction efficiency of 78, 78, and 81 % for odor, H_2S, and NH_3, respectively. Each pollutant needs a minimum EBRT depending on its loading rate and media moisture content. EBRTs between 4 and 10 s should be sufficient for a biofilter designed to control odors and VOCs from agricultural sites provided the moisture content is controlled adequately (Chen and Hoff 2009). Higher loading rates and lower media moisture content generally need a longer EBRT for an effective removal.
5. Moisture content: the humidity of air stream is important for maintaining the moisture content of the biofilter media and the biofilm. The media moisture content has been verified as a critical factor influencing biofilter performance as it supports the microbial population. Contaminated air streams introduced to the biofilter are usually pumped through a humidifier prior to entering the biofilter to reach a relative humidity greater than 95 % (Abd El-Bary 2003). A range of 40–65 % is believed suitable for media commonly used in agriculture, such as compost-based and wood chip-alone media (Chen and Hoff 2009). In addition to humidifying the airflow, sprinkler systems are frequently installed inside the biofilter that can be controlled to maintain suitable media moisture.
6. Temperature: the biofiltration appears to be an effective treatment process in the temperature range of 25–35 °C (Lu and Lin 1999). The temperature range from 20 to 40 °C has been recommended, with 35 °C believed optimal for biofilter operation. However, a wider temperature ranging from 4 to 40 °C has also shown high reduction efficiencies (Chen and Hoff 2009).
7. Biofilter pH and nutrients: the byproducts of microbial degradation are organic acids. In order to maintain the pH of the biofilter media around neutral, i.e., a pH around 7, buffering material may be added to the biofilter media (Abd El-Bary 2003). Nutrients should be kept in mind when biofilters are designed and operated. There are no guidelines identifying the amount of available nutrients needed in biofilters (Chen and Hoff 2009). Various nutrients supplied by compost-based media, which are commonly used in agriculture, in addition to the nutrients from exhaust air make supplemental nutrients unnecessary.
8. Pressure drop: the pressure drop is closely related to media type, media depth, moisture content in the media, media pore size, and air flow rate through the media. The pressure drop across the biofilter should be minimized since an increase in pressure drop requires higher blowing power and can result in air channeling through the media. The pressure drop through biofilters should be limited to no more than 50 Pa (Chen and Hoff 2009). Increased moisture and

decreased pore size result in increased pressure drop. Therefore, the media selection and moisturizing are critical to biofilter performance and energy efficiency.

9. Maintenance: the operation and maintenance of the biofilter would require weekly site visits during initiation of operations. However, after acclimatization and all system problems are resolved, the frequency of site visits could be reduced to be biweekly or monthly.

3.7.2 *Media of Biofilter*

Generally, the media should be capable of providing nutrients to the microorganisms and minimizing pressure drop. There are some criteria for choosing an optimum biofilter medium. In order to operate effectively in recycle systems, the media used in attached growth systems must have a relatively high specific surface area, i.e., surface area per unit volume, and an appreciable voids ratio. The specific surface area controls the amount of bacterial growth that can be supported in a unit volume. The voids ratio is critical for adequate performance of the system. The voids ratio characterizes how much space is provided for the fluid to pass through the media in close contact with the biofilm.

The media used in the biofilters must be inert, noncompressible and not biologically degradable. The media carries the biofilm where the bacterial growth occurs, and therefore, the media components are called biocarriers. A great variety of media materials have been verified suitable for biofilters. The media used in biofilters can include peat, heather, bark, composted sewage sludge, granular carbon, or other suitable materials. The randomly packed media can be sand, crushed rock, river gravel, plastic biocarriers, ceramic material shaped as small beads or larger spheres, rings, or saddles. However, the practical application in agricultural facilities and factors such as cost and local availability must be considered. Chen and Hoff (2009) stated that the mixture of compost and wood chips (ratio of 30–70 by weight) has been recommended as one of the better choices. Wood chips alone are another good option assuming enough bacteria and nutrients exist in the exhaust air. If not, inoculation can be achieved with compost and soil as well as activated sludge. Figure 3.9 shows hardwood and cedar media. The structured media can be crossed stacks of redwood slats, or plastic blocks composed of corrugated tubes or plates. Pressurized-bead filters and fluidized-bed filters use a finely graded plastic or sand media with average equivalent diameters generally from 1 to 3 mm and from 0.1 to 1.5 mm, respectively. Chen and Hoff (2009) stated that degradation of biofilter media, along with degradation of pollutants, is unavoidable. Biofilter life can be increased using a higher ratio of hardly degraded or nondegraded media materials. Remixing of media can increase biofilter life. A reasonable media lifespan of 3 years up to 5 years can be expected without causing a large pressure drop.

Fig. 3.9 Hardwood (HW) and western cedar (WC) media (Chen 2008)

Fig. 3.10 Wood chips media (*Source* Hartmann Filter GmbH & Co.KG.)

The shape of the media and its dimension is vital. The media of biofilters are characterized in terms of their key properties which include pH, total organic carbon (TOC), nitrogen and moisture content, oxygen uptake rates, and heterotrophic and fungal plate counts. Based on the physical characteristics of the various packing media, soft-wood chips over 75 mm screen size (Fig. 3.10) appeared to be the most promising because they provide one of the lowest pressure drops, the lowest coefficient of variation for air distribution across the pacing and are the least compressible (Phillips et al. 1995). All of these three physical factors are important for efficient minimum cost operation of biofilters in livestock buildings. The surface area of the media must be given a special consideration. The nitrification capacity of biological filters is largely dependent upon the total surface area available for biological growth and the efficiency of the area utilization, i.e., greater gas removal capacity is resulted from an increase in biofilm surface area (Summerfelt and Cleasby 1996). The performance of biofilter media should be evaluated continuously. The biofilter media characteristics that affect the performance of biofilters include pH, moisture content, oxygen uptake rates, total organic carbon (TOC), and microbial plate counts including total heterotrophs and fungi (Cardenas-Gonzalez et al. 1999).

Biofilter function can be impaired if a biofilter has too little oxygen or is overloaded with solids, biochemical oxygen demand (BOD), or ammonia. Increased loading can reduce the capacity of the biofilter to complete the two-step bacterial conversion of ammonia to nitrate, because increasing the loading increase the competition for space oxygen among heterotrophic organisms and nitrobacter, and nitrosomonas. Nitrobacter require the nitrite produced by nitrosomonas and tend to be located toward the outlet end of the biofilter. If space and oxygen become limiting when loading is increased, nitrobacter are generally the first to be displaced, as they are the last in the line of microbial consumers, but nitrosomonas can also be displaced or suffer from oxygen-limiting conditions. Conditions that cause a reduction in the relative amount of nitrobacter compared with nitrosomonas will cause the nitrite concentration across the biofilter to increase, which can become toxic (Abd El-Bary 2003). Two techniques have been applied, sometimes simultaneously, in order to solve the problem of insufficient ammonia and nitrite removal in large recirculation systems. The first technique is that the biofilter is sized to allow larger surface area for microbial treatment with consideration of oxygen requirements, and consequently the biofilter has both the space and oxygen capacity to handle higher odorous compounds and pollutants' loading rates. The second technique is that ozone is added to the system to oxidize excess nitrite to nitrate.

The organic oxidation capacity of biological filters is largely dependent upon the total surface area available for biological growth and the efficiency of the area utilization. Ideally, increasing the surface area of the media will result in a corresponding increase in harmful and malodorous gas removal capacity. The efficiency of nitrification per unit surface area is depending upon the accessibility of the media surface, the mass transfer rate into the biofilm, the growth phase of the biofilm (lag, log, stationary, or death phase), and the competition with heterotrophic microbes for space and oxygen. Nitrification rates reported for different filter types range from 0.04 to 0.78 g total ammonia nitrogen (TAN) removed per day per square meter of surface area, i.e., 0.04–0.78 g TAN/m^2 day (CIGR 1999).

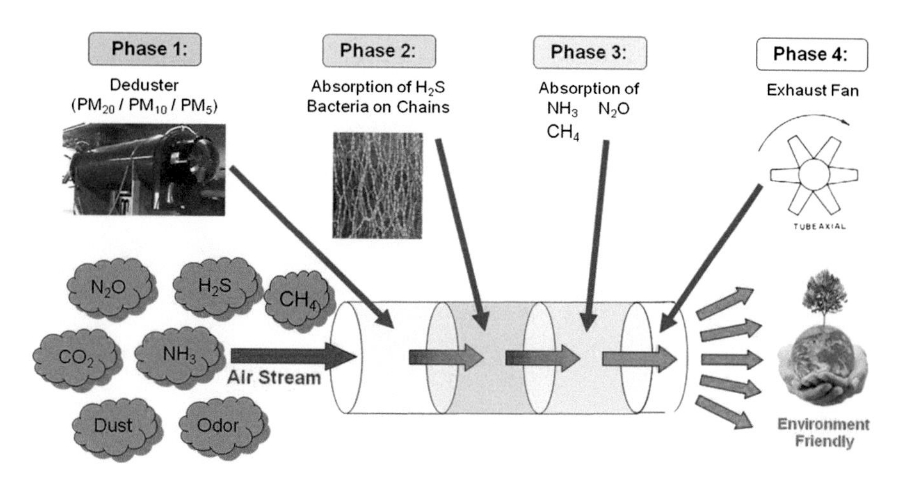

Fig. 3.11 Multiphase biological–chemical filter (Samer 2014)

3.7.3 Recent Advancements

Samer (2014) proposed a technique for climate adaptation that is able to reducing gas, odor, and dust emissions from livestock buildings; especially cattle, broiler, and pig housing, through prototyping and implementing a biological–chemical filter (BioChemiFilter). Nanotechnology and laser radiation is implemented to enhance the BioChemiFilter efficiency as follows: (1) A novel prototype of BioChemiFilter (BCF) was developed, and the proposed BCF consists of four phases/stages: the first phase is a deduster which catches dust (PM_5, PM_{10}, and PM_{20}); the second phase is a desulfurization unit which consists of chains where bacterial colonies of hydrogen sulfide (H_2S) scavengers are cultured, and this unit catches H_2S and decomposes it into harmless sulfa and water; in the third phase ammonia (NH_3) and nitrous oxide (N_2O) are caught by nitrosomonas which use them as energy source for themselves but they release nitrite as by-product; the nitrite will be transformed to nitrate by nitrobacter and the nitrate will be sequestered by sodium at the end of this phase to produce sodium nitrate which is a food preservative. (2) Nanoparticles are prepared and then used to treat the bacterial colonies where the nanoparticles replace the relevant heavy metal found in the structure of the co-enzyme; this process increases the bacterial activity. (3) Laser radiation is implemented to treat the bacterial colonies which stimulates cell division; consequently, the bacterial cell count increases. The calculations showed that implementing the proposed BCF (Fig. 3.11) can reduce the emissions from 4059194 to 170820 kg CO_2 equivalent per year from a typical dairy barn housing 150 cows. Consequently, an amount of 3888374 kg CO_2 equivalent can be saved which represents 95.8 % of the total greenhouse gases and carbon dioxide global warming potential.

Chapter 4
Perspective

The environmental legislations in Europe impose the following: (1) establishment, operation, and substantial alteration of livestock installations are subject to license; (2) sort and complexity of the licensing procedure depend on the location/site, species, and number of animals and the possible environmental impact (odor, ammonia, dust and nitrogen); (3) building permit is obligatory, where the initiation of an immission licensing procedure depends on the planned number of animal places; and (4) the permit according to an immission licensing procedure includes further official decisions related to the planned installation, in particular the building permit. Therefore, several studies focused on assessing minimum distances to prevent odor nuisance and harmful environmental impacts, setting emission, and immission limit values. Consequently, technical requirements and standards, best available techniques, and good agricultural practices are obligatory.

The emissions abatement techniques (Fig. 4.1) and, therefore, the mitigation strategies are facing several problems, which can be summarized as follows: (1) sophisticated quantification of emissions from the total production chain (animal housing, manure storage, and manure application); (2) the described measures are only single measures; however, the actual emissions from a livestock building can differ quite clearly; (3) potential reduction effect of a measure can be overlaid or compensated by other influencing factors; (4) large variability in measurements

M. Samer, *Abatement Techniques for Reducing Emissions from Livestock Buildings*, SpringerBriefs in Environmental Science,
DOI 10.1007/978-3-319-28838-3_4

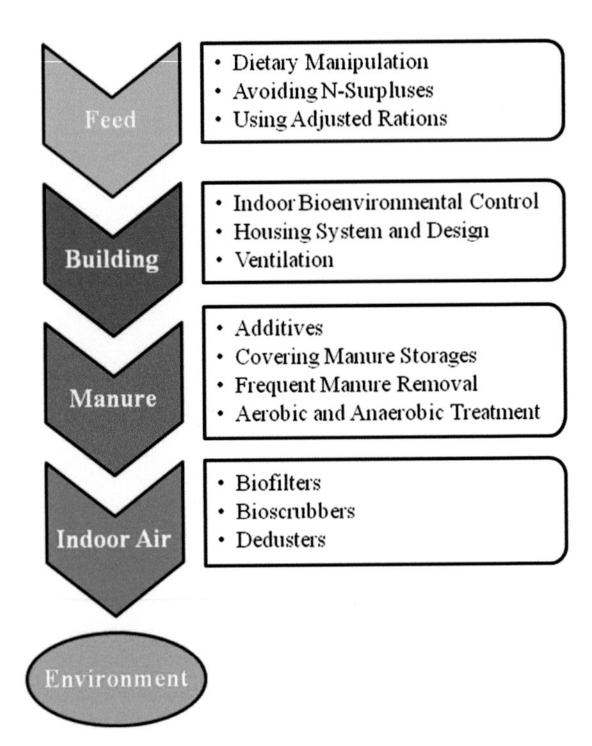

Fig. 4.1 Outline of emissions abatement techniques

from different housing systems and designs is adding to the complications in estimating emission factors and developing reduction strategies; and (5) conflict of interests between animal welfare, economic and environmental aspects.

Chapter 5
Summary and Conclusions

According to the issues raised in this study, the following can be further concluded:

1. The environmental legislations must regulate the establishment and operation of livestock farms through licensing procedures. Licenses are conferred upon farms based on their commitment to technical requirements and standards, best available techniques, and good agricultural practices which prevent negative environmental impacts.
2. Developing emissions abatement techniques to reduce emissions from livestock farms should be preceded with developing national emissions inventory which will provide a database on emission sources from which potentials of effective mitigation strategies can be drawn and relevant emissions abatement techniques can be applied.
3. It is crucial to consider the manure management systems inside the livestock buildings (slatted floor, manure scrapper, litter-based system…), floor type, building design, and ventilation system where these parameters have discernible influence on the emissions flow rates. Therefore, each farm should be treated as special case, where some farms may have applied emissions abatement techniques and other farms may have not, and therefore farm survey is necessary to be taken into consideration when inventorying the emissions and before developing mitigation strategies to reduce greenhouse gases and ammonia emissions from livestock manure.
4. Understanding of the development, release, and spreading processes of dust, odor, CH_4, CO_2, N_2O, NO, and NH_3 into the indoor air will contribute to development of emissions abatement techniques and housing designs, and will contribute to the reduction of odor, dust and gaseous emissions.
5. It is crucial to determine the emission rates, fluxes, and factors before and after deploying an emission abatement technique, where this should be considered in amending the present emissions inventory of the geographic area where the considered abatement technique has been deployed.

M. Samer, *Abatement Techniques for Reducing Emissions from Livestock Buildings*, SpringerBriefs in Environmental Science,
DOI 10.1007/978-3-319-28838-3_5

6. There are several techniques to reduce greenhouse gases and ammonia emissions such as dietary manipulation, manure treatment, additives utilization, covering manure storages, reducing the emission source surfaces, decreasing temperature and air velocity near the source, and minimizing volumetric airflow rates throughout the livestock buildings.
7. There are several means for reduction of dust in/from livestock houses, such as dust suppression with spraying oil and/or water, ionization, cleaning the air by oxidization using chemical compounds called "oxidants", use of windbreaks, and implementation of dedusters, bioscrubbers, and filters as either dry or wet.
8. There are a variety of approaches to removing odorous compounds from effluent streams: biofiltration, absorption, adsorption, condensation, chemical and/or biochemical reaction, incineration, and selective diffusion through a membrane. Biofilter technique is an effective technique for the elimination of malodorous gas emissions (odorous compounds, hydrogen sulfide, and ammonia emissions) and volatile organic compounds (VOCs).
9. The implementation of an emissions abatement technique for reducing a particular gas may change the concentrations of other pollutants (e.g., dust, CO_2, H_2S, etc...), may have an effect on welfare and productivity of the animals, should be feasible and have low investment and operating costs, and should provide a safe environment for farmers.

References

Aarnink AJA, Ellen HH (2007) Processes and factors affecting dust emissions from livestock production. In: International conference on How to improve air quality, Maastricht, the Netherlands (April 23–24)

Aarnink A, van Quwerkerk E, Verstegen M (1992) A mathematical model for estimating the amount and composition of slurry from fattening pigs. Livestock Prod Sci 31:133–147

Abd El-Bary KM (2003) Engineering and environmental studies on ammonia emitted from poultry houses. PhD Dissertation, Cairo University, Giza, Egypt

Adviento-Borbe MAA, Wheeler EF, Brown NE, Topper PA, Graves RE, Ishler VA, Varga GA (2010) Ammonia and greenhouse gas flux from manure in freestall barn with dairy cows on precision fed rations. Trans ASABE 53(4):1251–1266

Al-Kanani T, Akochi E, Mackenzie AF, Alli I, Barrington S (1992) Organic and inorganic amendments to reduce ammonia losses from liquid hog manure. J Environ Qual 21:709–715

Amon B (1998) NH3-, N2O- und CH4- Emissionen aus der Festmistanbindehaltung für Milchvieh – Stall-Lagerung-Ausbringung (In German, NH3, N2O and CH4 emissions from solid manure in tie-stalls for dairy cattle stall, storage and application). VDI-MEG Script 331, ISSN-Nr. 0931-6264. PhD Dissertation, University of Natural Resources and Life Sciences, Vienna, Austria

Bartzanas T, Kittas C, Sapounas AA, Nikita-Martzopoulou C (2007) Analysis of airflow through experimental rural buildings: sensitivity to turbulence models. Biosyst Eng 97:229–239

Beck J, Burton C (1998) Manure treatment techniques in Europe—result of a EU Concerted Action. In: Proceedings of the international conference on agricultural engineering, AgEng 98, Oslo, Norway, pp 211–212 (Aug 24–27)

Berckmans D, Vranken E (2006) Monitoring, prediction, and control of the microenvironment. In: Munack A (ed) CIGR handbook of agricultural engineering, vol VI., Information technologyASAE, St. Joseph, MI, USA, pp 383–401

Berg W (1999) Technology assessment—livestock management. Anim Res Dev 50:98–109

Berg W (2003) Reducing ammonia emissions by combining covering and acidifying liquid manure. In: Proceedings of the third international conference on air pollution from agricultural operations, Raleigh, NC, USA, pp 174–182 (Oct 12–15)

Berg W, Hoernig G (1997) Emission reduction by acidification of slurry—investigations and assessment. In: Voormans JAM, Monteny GJ (eds) Proceedings of the international symposium on Ammonia and Odour control from animal production facilities, Vinkeloord, The Netherlands, pp 459–466 (Oct 6–10)

Berg W, Pazsiczki I (2003) Reducing emissions by combining slurry covering and acidifying. In: Proceedings of the international symposium on Gaseous and Odour emissions from animal production facilities, Horsens, Denmark, pp 460–468 (June 1–4)

Berg W, Samer M (2010) Emissions from manure tanks. In: Presented at the international DLG exhibition for animal husbandry and management (EuroTier), Hannover, Germany (Nov 16–19)

M. Samer, *Abatement Techniques for Reducing Emissions from Livestock Buildings*, SpringerBriefs in Environmental Science,
DOI 10.1007/978-3-319-28838-3

Berg W, Brunsch R, Pazsiczki I (2006) Greenhouse gas emissions from covered slurry compared with uncovered during storage. Agric Ecosyst Environ 112(2–3):129–134

Berg W, Türk M, Hellebrand J (2006b) Effects of acidifying liquid cattle manure with nitric or lactic acid on gaseous emissions. Workshop on Agricultural Air Quality—State of the Science. Potomac, pp 492–498 (5–8 June 2006)

Bjerg B, Sørensen LC (2008) Numerical simulation of airflow in livestock buildings with radial inlet. Agric Eng Int: CIGR J BC 06:015

Bjorneberg DL, Leytem AB, Westermann DT, Griffiths PR, Shao L, Pollard MJ (2009) Measurements of atmospheric ammonia, methane, and nitrous oxide at a concentrated dairy production facility in southern Idaho using open-path FTIR spectrometry. Trans ASABE 52(5):1749–1756

Blanes-Vidal V, Topper PA, Wheeler EF (2007) Validation of ammonia emissions from dairy cow manure estimated with a non-steady-state, recirculation flux chamber with whole building emissions. Trans ASABE 50(2):633–640

Brink C, Kroeze C, Klimont Z (2001) Ammonia abatement and its impact on emissions of nitrous oxide and methane - Part 2: application for Europe. Atmos Environ 35(36):6313–6325

Burt S (2004) Essential oils: their antibacterial properties and potential applications in foods—a review. Int J Food Micro 94:223–253

Burton CH, Turner C, Beck JAF, Martinez J, Martens W, Pahl O, Piccinini S, Svoboda I (2003) Manure management—treatment strategies for sustainable agriculture, 2nd edn, Silsoe Research Institute, Wrest Park. ISBN 0 9531282 6 1

Cai L, Koziel JA, Liang Y, Nguyen AT, Xin H (2007) Evaluation of zeolite for control of odorants emissions from simulated poultry manure storage. J Environ Qual 36:184–193

Capper JL (2012) Is the grass always greener? Comparing the environmental impact of conventional, natural and grass-fed beef production systems. Animals 2:127–143

Carew R (2010) Ammonia emissions from livestock industries in Canada: Feasibility of abatement strategies. Environ Pollut 158(8):2618–2626

Cardenas-Gonzalez B, Ergas SJ, Switzenbaum MS, Phillibert N (1999) Evaluation of full-scale biofilter media performance. Environ Prog 18(3):205–211

Castillo-Gonzalez H, Bruns MA (2005) Dissimilatory iron reduction and odor indicator abatement by biofilm communities in swine manure microcosms. Appl Environ Microbiol 71(9): 4972–4978

Chadwick D, Sommer S, Thorman R, Fangueiro D, Cardenas L, Amon B, Misselbrook T (2011) Manure management: implications for greenhouse gas emissions. Animal Feed Sci Technol 166–167(2011):514–531

Chen L (2008) Mitigating odors from animal facilities using biofilters. PhD Dissertation, Iowa State University, USA

Chen L, Hoff SJ (2009) Mitigating odors from agricultural facilities: a review of literature concerning biofilters. Appl Eng Agric 25(5):751–766

CIGR (1984) Climatization of animal houses. Working Group Report on: Climatization of animal houses. International Commission of Agricultural Engineering (CIGR), Scotland

CIGR (1992) Climatization of animal houses. 2nd report of Working Group Report on: Climatization of animal houses. Ghent, Belgium: International Commission of Agricultural Engineering (CIGR) and Centre for Climatization of Animal Houses

CIGR (1994) Aerial environment in animal housing: Concentrations in and emissions from farm buildings. Working Group Report Series number 94.1: Climatization and environmental control in animal housing. International Commission of Agricultural Engineering (CIGR)

CIGR (1999) Handbook of agricultural engineering. Volume II. Animal Production & Aquaculutral Engineering. The American Society of Agricultural Engineers. Niles Roads, St. Joseph, Michigan

Colbeck I, Mackenzie AR (1994) Air pollution by photochemical oxidants. Air quality monographs, Vol. 1, ed by: Elsevier science B.V., ISBN 0-444-88542-0

Cottle DJ, Nolan JV, Wiedemann SG (2011) Ruminant enteric methane mitigation: a review. Animal Prod Sci 51(6):491–514

Dämmgen U, Erisman JW (2002) Transmission und Deposition von Ammoniak und Ammonium. In: Emissionen der Tierhaltung—Grundlagen, Wirkungen, Minderungsmaßnahmen (In German, Transmission and deposition of ammonia and ammonium. In: Emissions of livestock - basics, effects, mitigation measures). KTBL/UBA-Symposium, December 3–5, 2001, KTBL-Script 406, pp 50 – 62, ISBN 3-7843-2143-7, The German Association for Technology and Structures in Agriculture (KTBL), Darmstadt, Germany

Dämmgen U, Erisman JW (2006) Emission, Ausbreitung und Immission von Ammoniak und Ammonium – Übersicht über den gegenwärtigen Stand des Wissens. In: Emissionen der Tierhaltung (Emission, transmission and immission of ammonia and ammonium—Overview of the current state of the art. In: Emissions of livestock). KTBL Conference, December 5–7, 2006, KTBL-Script 449, pp 65–78, The German Association for Technology and Structures in Agriculture (KTBL), Darmstadt, Germany

Dekker SEM, Aarnink AJA, de Boer IJM, Groot Koerkamp PWG (2011) Emissions of ammonia, nitrous oxide, and methane from aviaries with organic laying hen husbandry. Biosyst Eng 110(2):123–133

DFG (2006) MAK-BAT-Werte-Liste 2006 (In German, List of Values for Best Available Techniques). Senatskommission zur Prüfung gesundheitsschädlicher Arbeitsstoffe (Commission for the Investigation of Health Hazards of Chemical Compounds), Mitteilung 42. Deutsche Forschungsgemeinschaft (German Research Council), Wiley-VCH Verlagsgesellschaft mbh, Weinheim, Deutschland

ECETOC (1994) Ammonia emissions to in Western Europe. Technical Report 62. European Centre for Ecotoxicology and Toxicology of Chemicals, Brussels

EMEP/EEA (2009) EMEP/EEA air pollutant emission inventory guidebook: Animal husbandry and manure management. The European Environment Agency (EEA) and the Cooperative programme for monitoring and evaluation of the long-range transmission of air pollutants in Europe (EMEP). http://eea.europa.eu/emep-eea-guidebook

Eniola B, Perschbacher-Buser ZL, Caraway E, Ghosh N, Olse M, Parker D (2006) Odor control in waste management lagoons via reduction of p-cresol using horseradish peroxidase. In: Proceedings of ASABE Annual Intl Meeting Oregon, p 7. July 2006, Paper No: 064044

European Commission (2003) Integrated Pollution Prevention and Control (IPPC): Intensive Rearing of Poultry and Pigs. Reference Document on Best Available Techniques for Intensive Rearing of Poultry and Pigs

Executive Body for the Convention on Long-Range Transboundary Air Pollution (2007a) Methods and Procedures for the Technical Review of Air Pollutant Emission Inventories Reported under the Convention and its Protocols. Economic Commission for Europe, Economic and Social Council, United Nations

Executive Body for the Convention on Long-Range Transboundary Air Pollution (2007b) Guidance Document on Control Techniques for Preventing and Abating Emissions of Ammonia. Economic Commission for Europe, Economic and Social Council, United Nations

FAO (2006) Livestock's role in climate change and air pollution. ftp://ftp.fao.org/docrep/fao/010/a0701e/A0701E03.pdf. Accessed Dec 2010

Filipy J, Rumburg B, Mount G, Westberg H, Lamb B (2006) Identification and quantification of volatile organic compounds from a dairy. Atmos Environ 40:1480–1494

Gay SW, Schmidt DR, Clanton CJ, Janni KA, Jacobson LD, Weisberg S (2003) Odor, total reduced sulfur, and ammonia emissions from animal housing facilities and manure storage units in Minnesota. Trans ASAE 19(3):347–360

Giltrap DL, Singh J, Saggar S, Zaman M (2010) A preliminary study to model the effects of a nitrification inhibitor on nitrous oxide emissions from urine-amended pasture. Agric Ecosyst Environ 136(3–4):310–317

Govere EM, Tonegawa M, Bruns MA, Wheeler EF, Kephart KB, Voigt JW, Dec J (2007) Using minced horseradish roots and peroxides for the deodorization of swine manure: a pilot scale study. Bioresour Technol 98:1191–1198

Gustafsson G (1997) Investigations of factors affecting air pollutants in animal houses. Ann Agric Environ Med 4:203–215

Haeussermann A, Hartung E, Gallmann E, Jungbluth T (2006) Influence of season, ventilation strategy, and slurry removal on methane emissions from pig houses. Agric, Ecosyst Environ 112(2006):115–121

Hagenkamp-Korth F, Haeussermann A, Hartung E, Reinhardt-Hanisch A (2015) Reduction of ammonia emissions from dairy manure using novel urease inhibitor formulations under laboratory conditions. Biosyst Eng 130(2015):43–51

Hartmann Filter GmbH & Co.KG., Glasebachstraße 30, 33165 Lichtenau, Germany. http://www.hartmann-filter.de/. Accessed 6 Jan 2015

Hartung E (1995) Entwicklung einer Messmethode und Grundlagenuntersuchung zur Ammoniakfreisetzung aus Flüssigmist (In German, Development of a measurement method and Basic research on ammonia release from liquid manure). VDI-MEG Script 275, ISSN-Nr. 0931-6264, PhD Dissertation, University of Hohenheim, Stuttgart, Germany

Hartung J, Phillips VR (1994) Control of gaseous emissions from livestock buildings and manure stores. J Agric Eng Res 57:173–189

Hartung J, Saleh M (2007) Composition of dust and effects on animals. International interdisciplinary conference. Particulate matter in and from agriculture, September 3-4, 2007, Braunschweig, Germany

Hellickson MA, Walker JN (1983) Ventilation of agricultural structures. ASAE, St. Joseph, Mich

Hendriks JGL, Vrielink MGM (1997) Reducing ammonia emission from pig houses by adding or producing organic acids in pig slurry. In: Voormans JAM, Monteny GJ (eds) Proceedings of the international symposium on ammonia and odour control from animal production facilities, Vinkeloord, The Netherlands, pp 493–501 (Oct 6–10)

Hessel EF, Van den Weghe HFA (2007) Airborne Dust (PM10) concentration in broiler houses as a function of fattening day, time of the day, and indoor light. International Conference on How to improve air quality, April 23 - 24, 2007, Maastricht, The Netherlands

Husted S, Jensen LS, Jorgensen SS (1991) Reducing ammonia loss from cattle slurry by the used of acidifying additives: the role of the buffer system. J Sci Food Agric 57:335–349

Hutchings NJ, Sommer SG, Andersen JM, Asman WAH (2001) A detailed ammonia emission inventory for Denmark. Atmos Environ 35(11):1959–1968

Illinois Odor and Nutrient Control Proving Center, University of Illinois at Urbana-Champaign, USA. http://age-web.age.uiuc.edu/bee/RESEARCH/proving-center/proving-center.html. Accessed 9 Jan 2015

international symposium, 21–23 May (Louisville, Kentucky, USA) ASAE Publication Number 701P0201, Stowell RR, Bucklin R, Bottcher RW (eds) pp 596–603.

IPCC (2006) Intergovernmental panel on climate change, guidelines for national greenhouse gas inventories. Volume 4: Agriculture, Forestry and Other Land Use. Chapter 10: Emissions from Livestock and Manure Management

IPCC (2007a) Climate Change 2007: Mitigation. Contribution of Working Group III to the 4th Assessment Report of the Intergovernmental Panel on Climate Change. Metz B, Davidson OR, Bosch PR, Dave R, Meyer LA (eds) Cambridge University Press, Cambridge, United Kingdom and New York

IPCC (2007b) Climate change 2007: climate change 2007: the physical science basis, Technical Summary

IPCC (2007) Climate change 2007. The Physical Science Basis, Summary for Policymakers

Janzen HH, Desjardins RL, Rochette P, Boehm M, Worth D (2008) Better farming better air. Agriculture and Agri-food Canada, Ottawa

Jensen FH (1974) Geruchsminderung durch Umwälzlüftung (In German, Odor reduction by recirculating ventilation). KTBL-Script 186, The German Association for Technology and Structures in Agriculture (KTBL), Darmstadt-Kranichstein, Germany

Jungbluth T, Buscher W (1996) Reduction of ammonia emissions from piggeries. ASAE Paper No. 96–4091. The ASAE Annual International Meeting, July 17 (1996) Phoenix. Arizona, USA

Keck M (1997) Beeinflussung von Raumluftqualität und Ammoniakemissionen aus der Schweinehaltung durch verfahrenstechnische Maßnahmen (In German, Influencing the indoor air quality and ammonia emissions from pig housing through procedural measures). VDI-MEG Script 299, ISSN-Nr. 0931-6264, PhD Dissertation, University of Hohenheim, Stuttgart

Kroodsma W, Ogink NWM (1997) Volatile emissions from cow cubicle houses and its reduction by immersion of the slats with acidified slurry. In: Voormans JAM, Monteny GJ (eds) Proceedings of the international symposium on ammonia and odour control from animal production facilities, Vinkeloord, The Netherlands, pp 475–483 (Oct 6–10)

KTBL (2006) Handhabung der TA Luft bei Tierhaltungsanlagen: Ein Wegweiser für die Praxis (In German, Handling of the Clean Air Act in livestock farming systems: A guide for the practice). The German Association for Technology and Structures in Agriculture (KTBL), KTBL-Script, Band 447, 244 p, ISBN 978-3-939371-14-4, Darmstadt, Germany

Kuczynski T, Blanes-Vidal V, Li BM, Gates RS, Nääs IA, Moura DJ, Berckmans D, Banhazi TM (2011) Impact of global climate change on the health, welfare and productivity of intensively housed livestock. Int J Agric Biol Eng 4(2):1–22

Lay JJ, Li YY, Noike T (1997) Influences of pH and moisture content on the methane production in high-solids sludge digestion. Water Res 31(6):1518–1524

Lehninger AL (1977) Biochemie (In German, Biochemistry). Verlag Chemie, Weinheim, 2nd edn, ISBN 3-527-25688-1

Li H, Xin H, Burns RT (2006) Reduction of ammonia emission from stored poultry manure using additives: zeolite, Al^+ clear, Ferix-3, and PLT. ASAE Paper No. 064188, 2006 ASABE Annual International Meeting, Portland

Lim TT, Heber AJ, Ni J-Q, Gallien JX, Xin H (2003) Air quality measurements at a laying hen house: particulate matter concentrations and emissions. Proceedings of Air Pollution from Agricultural Operations III Conference, October 12-15, 2003 Research Triangle Park, North Carolina USA, ASAE Publication Number 701P1403, ed H Keener. pp 249-256

Lu C, Lin M-R (1999) Temperature effects of trickle-bed biofilter for treating BTEX vapors. J Environ Eng 125(8):775–779

Martinez J, Jolivet J, Guiziou F, Langeoire G (1997) Ammonia emissions from pig slurries: evaluation of acidification and the use of additives in reducing losses. In: Voormans JAM, McCrory DF, Hobbs PJ 2001 Additives to reduce ammonia and odor emissions from livestock wastes: a review. J Environ Qual 30:345–355

Melse RW, Van Der Werf AW (2005) Biofiltration for mitigation of methane emission from animal husbandry. Environ Sci Technol 39(14):5460–5468

Melse RW, Ogink NWM, Rulkens WH (2009) Overview of European and Netherlands' regulations on airborne emissions from intensive livestock production with a focus on the application of air scrubbers. Biosyst Eng 104(3):289–298

Merino P, Ramirez-Fanlo E, Arriaga H, del Hierro O, Artetxe A, Viguria M (2011) Regional inventory of methane and nitrous oxide emission from ruminant livestock in the Basque Country. Animal Feed Sci Technol 166–167(2011):628–640

Miller DN, Varel VH (2001) In vitro study of the biochemical origin and production limits of odorous compounds in cattle feedlots. J Animal Sci 79:2949–2956

Mitchell BW, Baumgartner JW (2007) Electrostatic space charge systems for dust reduction in animal housing. The 2007 ASAE Annual International Meeting, Chicago, Paper Number 074176, doi:10.13031/2013.23274

Mitchell BW, Richardson LJ, Wilson JL, Hofacre CL (2004) Application of an electrostatic space charge system for dust, ammonia and pathogen reduction in a broiler breeder house. Appl Eng Agric 20(1):87–93

Mölter L, Schmidt M (2007) Advantages and limits of aerosol spectrometers for the particle size and particle quantity determination stables and air exhaust ducts. International interdisciplinary conference on Particulate matter in and from agriculture, September 3-4, 2007, Braunschweig, Germany

Monteny GJ (2000) Modelling of ammonia emissions from dairy cow houses. PhD Dissertation, 156 p, Wageningen University, The Netherlands

Morsing S, Strom JS, Zhang G, Kai P (2008) Scale model experiments to determine the effects of internal airflow and floor design on gaseous emissions from animal houses. Biosyst Eng 99: 99–104

Mostafa E (2008) Improvement of air quality in laying hens barn using different particle separation techniques. PhD diss. University of Bonn, Bonn

Mostafa E, Buescher W (2011) Indoor air quality improvement from particle matters for laying hen poultry houses. Biosyst Eng 109(1):22–36

Mostafa E (2012) Air-Polluted with Particulate Matters from Livestock Buildings, Air Quality—New Perspective, Lopez G (ed), ISBN: 978-953-51-0674-6, InTech, doi:10.5772/45766

Nannen C (2005) Mikroskopische Untersuchung von Luftgetragenen Partikeln in Schweinemastställen (In German, Microscopic examination of airborne particles in pig barns). M.Sc. Thesis, Institute for Agricultural Engineering, University of Bonn

Ngwabie NM, Jeppsson K-H, Nimmermark S, Swensson C, Gustafsson G (2009) Multi-location measurements of greenhouse gases and emission rates of methane and ammonia from a naturally-ventilated barn for dairy cows. Biosyst Eng 103:68–77

Ngwabie NM, Jeppsson K-H, Nimmermark S, Gustafsson G (2011) Effects of animal and climate parameters on gas emissions from a barn for fattening pigs. Appl Eng Agric 27(6):1027–1037

Nicolai RE, Janni KA (1999) Effect of biofilter retention time on emissions from diary, swine, and poultry buildings. ASAE Paper No 99-4149. The ASAE CSAE Annual International Meeting, Toronto (July 18–22)

Nicolai RE, Lefers RM (2006) Biofilters used to reduce emissions from livestock housing: A literature review. In Workshop on Agricultural Air Quality, pp. 952-960, Washington, D.C.: Dept of Communication Services, North Carolina State University, Raleigh

Nicolai RE, Thaler R (2007) Vertical biofilter construction and performance. ASABE Publication No. 701P0907cd. St. Joseph, Mich.: ASABE

Nicolai RE, Janni KA (1999) Effect of biofilter retention time on emissions from diary, swine, and poultry buildings. ASAE Paper No. 99-4149. The ASAE CSAE Annual International Meeting, Toronto (July 18–22)

Oenema O, Velthof GL (1993) Denitrification in nitric-acid-treated cattle slurry during storage. Neth J Agric Sci 41:63–80

O'Neill DH, Phillips VR (1992) A review of the control of odour nuisance from livestock buildings. Part 3: Properties of the odorous substances which have been identified in livestock wastes or in the air around them. J Agric Eng Res 53:23–50

Pain BF, Thompson RB, Rees YJ, Skinner JH (1990) Reducing gaseous losses of nitrogen from cattle slurry applied to grassland by the use of additives. J Sci Food Agric 50:141–153

Pedersen S, Nonnenmann M, Rautiainen R, Demmers TGM, Banhazi T, Lyngbye M (2000) Dust in pig buildings. J Agric Saf Health 6(4):261–274

Pedersen P (2003) Reduction of gaseous emissions from pig houses by adding sulphuric acid to the slurry. In: Proceedings of the international symposium on Gaseous and Odour emissions from animal production facilities, Horsens, Denmark, pp 257–263 (June 1–4)

Pereira J, Fangueiro D, Misselbrook T, Chadwick D, Coutinho J, Trindade H (2011) Ammonia and greenhouse gas emissions from slatted and solid floors in dairy cattle houses: a scale model study. Biosyst Eng 109:148–157

Phillips VR, Scotford IM, White RP, Hartshorn RL (1995) Minimum-cost biofilters for reducing odours and other aerial emissions from livestock buildings: Part 1, Basic airflow aspects. J Agric Eng Res 62(3):203–214

Rabaud NE, Ebeler SE, Ashbaugh LL, Flocchini RG (2003) Characterization and quantification of odorous and non-odorous volatile organic compounds near a commercial dairy in California. Atmos Environ 39:933–940

Rahman S, DeSutter T, Zhang Q (2011) Efficacy of a microbial additive in reducing odor, ammonia, and hydrogen sulfide emissions from farrowing-gestation swine operation. Agric Eng Int CIGR J 13(3):1–9

Reidy B, Rhim B, Menzi H (2008) A new Swiss inventory of ammonia emissions from agriculture based on a survey on farm and manure management and farm-specific model calculations. Atmos Environ 42(2008):3266–3276

Reidy B, Dammgen U, Dohler H, Eurich-Menden B, van Evert FK, Hutchings NJ, Luesink HH, Menzi H, Misselbrook TH, Monteny G-J, Webb J (2008) Comparison of models used for national agricultural ammonia emission inventories in Europe: Liquid manure systems. Atmos Environ 42(2008):3452–3464

Reidy B, Webb J, Misselbrook TH, Menzi H, Luesink HH, Hutchings NJ, Eurich-Menden B, Dohler H, Dammgen U (2009) Comparison of models used for national agricultural ammonia emission inventories in Europe: Litter-based manure systems. Atmos Environ 43(2009): 1632–1640

Reinhardt-Hanisch A (2008) Grundlagenuntersuchungen zur Wirkung neuartiger Ureaseinhibitoren in der Nutztierhaltung (Basic research on the effects of novel urease inhibitors in animal housing). PhD Dissertation, University of Hohenheim, Stuttgart

Ritter WF, Collins NE, Eastburn RP (1975) Chemical treatment of liquid dairy manure to reduce malodours. In: Managing livestock manure. Proceedings of the 3rd International Symposium on livestock manure. Publication Proc-275. American Society of Agricultural Engineers, St. Joseph, MI, pp 381–384

Ritz CW, Mitchell BW, Fairchild BD, Czarick M III, Worley JW (2006) Improving in-house air quality in broiler production facilities using an electrostatic space charge system. J Appl Poult Res 15(2):333–340

Robert WB (2001) An environmental nuisance: odour concentration and transported by dust. Chem Senses 26:327–331

Samer M (2010) A software program for planning and designing biogas plants. Trans ASABE 53(4):1277–1285

Samer M (2011) How to construct manure storages and handling systems? IST Trans Biosyst Agric Eng 1(1):1–7

Samer M (2011) Seasonal variations of gaseous emissions from a naturally ventilated dairy barn. Misr J Agric Eng 28(4):1162–1177

Samer M (2012) Effect of airflow profile on reducing heat stress, enhancing air distribution and diluting gaseous concentrations in dairy barns. Misr J Agric Eng 29(2):837–856

Samer M (2012b) Reconstruction of old gutter-connected dairy barns: A case study. Proceedings of the 2012 American Society of Agricultural and Biological Engineers (ASABE) Annual International Meeting, 29.07-01.08.2012, Paper No. 121341061, 2012(7), pp 5401–5418, Dallas, Texas

Samer M (2012c) Biogas Plant Constructions, pp 343–368. In: Kumar BS (ed) ISBN 978-953-51-0204-5. Rijeka, Croatia: InTech. doi:10.5772/31887

Samer M (2013) Emissions inventory of greenhouse gases and ammonia from livestock housing and manure management. Agric Eng Int CIGR J 15(3):29–54

Samer M (2013) Towards the implementation of the Green Building concept in agricultural buildings: a literature review. Agric Eng Int CIGR J 15(2):25–46

Samer M (2014) Implementation of nanotechnology and laser radiation to prototype a biological-chemical filter for reducing gas, odor and dust emissions from livestock buildings. Session: Measurement and Mitigation of Pollutants from Livestock and Poultry Housing. Proceedings of the 2014 American Society of Agricultural and Biological Engineers (ASABE) Annual International Meeting, a joint conference with the Canadian Society of Bioengineering (CSBE), 13-16.07.2014, Paper No. 141870677, Montreal, Quebec

Samer M (2015) GHG Emission from Livestock Manure and its Mitigation Strategies. In: Climate Change Impact on Livestock: Adaptation and Mitigation, Sejian V, Gaughan J, Baumgard L, Prasad C (eds), pp.321–346, ISBN 978-81-322-2264-4, Springer International, Germany

Samer M, Grimm H, Hatem M, Doluschitz R, Jungbluth T (2008a) Mathematical modeling and spark mapping for construction of aerobic treatment systems and their manure handling system. Proceedings of International Conference on Agricultural Engineering, Book of Abstracts p 28, EurAgEng, June 23-25, Hersonissos, Crete, Greece

Samer M, Grimm H, Hatem M, Doluschitz R, Jungbluth T (2008) Mathematical modeling and spark mapping of dairy farmstead layout in hot climates. Misr J Agric Eng 25(3):1026–1040

Samer M, Loebsin C, von Bobrutzki K, Fiedler M, Ammon C, Berg W, Sanftleben P, Brunsch R (2011) A computer program for monitoring and controlling ultrasonic anemometers for aerodynamic measurements in animal buildings. Comput Electron Agric 79(1):1–12

Samer M, Loebsin C, Fiedler M, Ammon C, Berg W, Sanftleben P, Brunsch R (2011) Heat balance and tracer gas technique for airflow rates measurement and gaseous emissions quantification in naturally ventilated livestock buildings. Energy Build 43(12):3718–3728

Samer M, Fiedler M, Müller H-J, Gläser M, Ammon C, Berg W, Sanftleben P, Brunsch R (2011) Winter measurements of air exchange rates using tracer gas technique and quantification of gaseous emissions from a naturally ventilated dairy barn. Appl Eng Agric 27(6):1015–1025

Samer M, Müller H-J, Fiedler M, Ammon C, Gläser M, Berg W, Sanftleben P, Brunsch R (2011) Developing the 85Kr tracer gas technique for air exchange rate measurements in naturally ventilated animal buildings. Biosyst Eng 109(4):276–287

Samer M, Berg W, Müller H-J, Fiedler M, Gläser M, Ammon C, Sanftleben P, Brunsch R (2011) Radioactive 85Kr and CO2-balance for ventilation rate measurements and gaseous emissions quantification through naturally ventilated barns. Trans ASABE 54(3):1137–1148

Samer M, Ammon C, Loebsin C, Fiedler M, Berg W, Sanftleben P, Brunsch R (2012a) Moisture balance and tracer gas technique for ventilation rates measurement and greenhouse gases and ammonia emissions quantification in naturally ventilated buildings. Build Environ 50(4):10–20

Samer M, Berg W, Fiedler M, von Bobrutzki K, Ammon C, Sanftleben P, Brunsch R (2012b) A comparative study among H2O-balance, heat balance, CO2-balance and radioactive tracer gas technique for airflow rates measurement in naturally ventilated dairy barns. Proceedings of the Ninth International Livestock Environment Symposium (ASABE ILES IX), Paper No. ILES12-0079, ASABE, Valencia, Spain (July 8–12)

Samer M, Abuarab ME (2014) Development of CO2-balance for ventilation rate measurements in naturally cross ventilated dairy barns. Trans ASABE 57(4):1255–1264

Samer M, Müller H-J, Fiedler M, Berg W, Brunsch R (2014) Measurement of ventilation rate in livestock buildings with radioactive tracer gas technique: theory and methodology. Indoor Built Environ 23(5):692–708

Samer M, Mostafa E, Hassan AM (2014b) Slurry treatment with food industry wastes for reducing methane, nitrous oxide and ammonia emissions. Misr Journal of Agricultural Engineering 31(4):1523–1548

Schilling G, Ansorge H, Borchmann W, Markgraf G, Peschke H (1989) Pflanzenernährung und Düngung: Teil II – Düngung (In German, Plant Nutrition and Fertilization: Part II - Fertilization). VEB Deutscher Landwirtschaftsverlag (German Agricultural Publisher), 2nd edn, ISBN 3-331-00014-0

Schneider B (1988) Computer-based continuous recording of heat, water vapor, and carbon dioxide production in livestock barns. University of Hohenheim, Germany, PhD diss

Schuurkes J, Mosello R (1988) The role of external ammonium inputs in freshwater acidification. Aquatic Sci Res Across Boundaries 50(1):71–86

Shah SB, Kolar P (2012) Evaluation of additive for reducing gaseous emissions from swine waste. Agric Eng Int CIGR J 14(2):10–20

Schmidt D, Janni KJ, Nicolai R (2004) Biofilter design information. BAEU-18. Biosystems and Agricultural Engineering Department, University of Minnesota. St. Paul, Minn. www.bbe.umn.edu/extens/aeu/baeu18.pdf. Accessed 22 April 2008

Seethapathy S, Górecki T, Li X (2008) Passive sampling in environmental analysis. J Chromatogr A 1184(1–2):234–253
Snell HGJ, Schwarz A (2003) Development of an efficient bioscrubber system for the reduction of emissions. ASAE annual international meeting, ASAE, Las Vegas, Vevada (27–30 July)
Snell HGJ, Seipelt F, van den Weghe HFA (2003) Ventilation rates and gaseous emissions from naturally ventilated dairy houses. Biosyst Eng 86:67–73
Sommer SG, Christensen BT, Nielsen NE, Schjorring JK (1993) Ammonia volatilization during storage of cattle and pig slurry—effect of surface cover. J Agric Sci 121:63–71
Sommer SG, Pedersen SO, Sogaard HT (2000) Greenhouse gas emissions from stored livestock slurry. J Environ Qual 29:744–751
Sommer SG, Petersen SO, Møller HB (2004) Algorithms for calculating methane and nitrous oxide emissions from manure management. Nutr Cycl Agroecosyst 69(2):143–154
Sommer SG, Zhang GQ, Bannink A, Chadwick D, Misselbrook T, Harrison R, Hutchings NJ, Menzi H, Monteny GJ, Ni JQ, Oenema O, Webb J (2006) Algorithms determining ammonia emission from buildings housing cattle and pigs and from manure stores. Adv Agron 89:261–335
SRES (2000) In: Nakicenovic N, Swart R (eds) Special Report on Emissions Scenarios (SRES). World Meteorological Organization, Geneva
Stevens RJ, Laughlin RJ, Frost JP (1989) Effect of acidification with sulphuric acid on the volatilization of ammonia from cow and pig slurries. Cambridge J Agric Sci 113:389–395
Summerfelt ST, Cleasby JL (1996) A review of hydraulics in fluidized-bed biological filters. Trans ASAE 39(3):1161–1173
Sun H, Stowell RR, Keener HM, Michel FC Jr (2002) Two-dimensional computational fluid dynamics (CFD) modeling of air velocity and ammonia distribution in a High-Rise™ hog building. Trans ASAE 45:1559–1568
Sunesson AL, Gullberg J, Blomquist G (2001) Airborne chemical compounds in dairy farms. J Environ Monit 23:210–216
Takai H, Pedersen S, Johnsen JO, Metz JHM, Koerkamp PWGG, Uenk GH, Phillips VR, Holden MR, Sneath RW, Short JL, White RP, Hartung J, Seedorf J, Schroder M, Linkert KH, Wathes CM (1998) Concentrations and emissions of airborne dust in livestock buildings in Northern Europe. J Agric Eng Res 70(1):59–77
Ullman JL, Mukhtar S, Lacey RE, Carey JB (2004) A review of literature concerning odors, ammonia, and dust from broiler production facilities: 4. Remedial management practices. J Appl Poult Res 13:521–531
UNECE (2007) Guidance document on control techniques for preventing and abating emissions of ammonia. United Nations Economic Commission for Europe (UNECE), Executive Body for the Convention on Long-Range Transboundary Air Pollution, Working Group on Strategies and Review, ECE/EB.AIR/WG.5/2007/13, Geneva, Switzerland
UNFCCC (2014) Global Warming Potentials, United Nations Framework Convention on Climate Change. http://unfccc.int/ghg_data/items/3825.php. Accessed 30 Dec 2014
Van Buggenhout S, Van Brecht A, Eren Ozcan S, Vranken E, Van Malcot W, Berckmans D (2009) Influence of sampling positions on accuracy of tracer gas measurements in ventilated spaces. Biosyst Eng 104(2009):216–223
Vergé XPC, Dyer JA, Worth DE, Smith WN, Desjardins RL, McConkey BG (2012) A greenhouse gas and soil carbon model for estimating the carbon footprint of livestock production in Canada. Animals 2:437–454
Von Bobrutzki K, Braban CF, Famulari D, Jones SK, Blackall T, Smith TEL et al (2010) Field inter-comparison of eleven atmospheric ammonia measurement techniques. Atmos Meas Tech 3:91–112
Von Bobrutzki K, Müller H-J, Scherer D (2011) Factors affecting the ammonia content in the air surrounding a broiler farm. Biosyst Eng 108:322–333
Wheeler EF, Topper PA, Brandt RC, Brown NE, Adviento-Borbe A, Thomas RS, Varga GA (2011) Multiple-chamber instrumentation development for comparing gas fluxes from biological materials. Appl Eng Agric 27(6):1049–1060

Wheeler EF, Adviento-Borbe MAA, Brandt RC, Topper PA, Topper DA, Elliott HA, Graves RE, Hristov AN, Ishler VA, Bruns MAV (2011) Amendments for mitigation of dairy manure ammonia and greenhouse gas emissions: preliminary screening. Agric Eng Int CIGR J 13(2):1–14

Wheeler EF, Adviento-Borbe MAA, Brandt RC, Topper PA, Topper DA, Elliott HA, Graves RE, Hristov AN, Ishler VA, Bruns MAV (2011) Evaluation of odor emissions from amended dairy manure: preliminary screening. Agric Eng Int CIGR J 13(2):15–29

Williams A (2003) Floating covers to reduce ammonia emissions from slurry. In: Proceedings of the international symposium on Gaseous and Odour emissions from animal production facilities, Horsens, Denmark, pp 283–291 (June 1–4)

Wright DW, Eaton DK, Nielse LT, Khurt FW, Koziel JA, Spinhirne JP, Parker DB (2004) Multidimensional gas chromatorgraphy-olfactometry for identification and prioritization of malodors from confined animal feeding operations. Proc. of ASAE/CSAE Annual International Meeting, 2004, Ontario, Canada

Yang Y, Tugna P (1999) Control of odorous air emissions at a municipal wastewater treatment plant using a biofilter. Paper No.105. Proceedings of the Annual Meeting & Exhibition of the Air & Waste Management Association, June 20–24, St. Louis, Missouri

Zhang Y (2000) Modeling and Sensitivity Analysis of Dust Particle Separation for Uniflow Dedusters. University of Illinois Urbana—Champaign, 1–13

Zhang R, McGarvey JA, Ma Y, Mitloehner FM (2008) Effects of anaerobic digestion and aerobic treatment on the reduction of gaseous emissions from dairy manure storages. Int J Agric Biol Eng 1(2):15–20

Zhang Y, Polakow J A, Wang X, Riskowski GL, Sun Y, Ford SE (2001) An aerodynamic deduster to reduce dust and gas emissions from ventilated livestock facilities. Proceedings of the sixth

Zhang G, Strom JS, Li B, Rom HB, Morsing S, Dahl P, Wang C (2005) Emission of ammonia and other contaminant gases from naturally ventilated dairy cattle buildings. Biosyst Eng 92:355–364

Zhang G, Zhang Y, Kim Y, Kim J, Liu L, Yu X, Teng X (2011) Field study on the impact of indoor air quality on broiler production. Indoor Built Environ 20(4):449–455

Zhu J, Jacobson LD (1999) Correlating microbes to major odorous compounds in swine manure. J Environ Qual 28:737–744

Zhu Z, Dong H, Tao X, Xin H (2005) Evaluation of airborne dust concentration and effectiveness of cooling fan with spraying misting systems in swine gestation houses. Proceedings of the seventh international symposium, May 18–20, 2005, Beijing, China, ASAE Publication Number 701P0205, Brown-Brandl T (ed) pp 224–229